The Kew

Temperate Plant Families Identification Handbook

Edited by
**Gemma Bramley,
Anna Trias-Blasi
& Richard Wilford**

Kew Publishing
Royal Botanic Gardens, Kew

First published in 2023 by the Royal Botanic Gardens, Kew, Richmond, Surrey, TW9 3AB, UK
www.kew.org

ISBN 978 1 84246 772 5
e-ISBN 978 1 84246 773 2

Distributed on behalf of the Royal Botanic Gardens, Kew in North America by the University of Chicago Press, 1427 East 60th Street, Chicago, IL 60637, USA.

British Library Cataloguing in Publication Data
A catalogue record for this book is available from the British Library.

Design: Nicola Thompson
Typesetting and page layout: Christine Beard
Copy-editing Sharon Whitehead
Proofreading: Ruth Linklater
Production management: Georgie Hills

Printed in the UK by Gomer Press Limited

For information or to purchase all Kew titles please visit shop.kew.org/kewbooksonline or email publishing@kew.org

Kew's mission is to understand and protect plants and fungi, for the wellbeing of people and the future of all life on Earth.

Kew receives approximately one third of its funding from Government through the Department for Environment, Food and Rural Affairs (Defra). All other funding needed to support Kew's vital work comes from members, foundations, donors and commercial activities, including book sales.

Contents

Family accounts (listed by Order)

Introduction

The Kew Temperate Plant Families Identification Handbook, featuring 100 plant families that are commonly encountered in the world's temperate regions, is designed to act as an identification guide and reference text. With full descriptions, shorter sections highlighting key spot characters and a comprehensive selection of images included for each family, it is a practical resource that facilitates the identification of plants in temperate zones; the handy size means it is portable for fieldwork and easy to carry around the workplace, be it laboratory, herbarium or office. The handbook's format is modelled on the companion volume, *The Kew Tropical Plant Families Identification Handbook* (2ⁿᵈ edition (revised)) (Utteridge & Bramley, 2020), a guide to the 100 most commonly encountered plant families in the world's tropical regions. Both volumes are aimed at conservation and environment professionals, but they are also suitable for graduate students, horticulturists, ethnobotanists, ecologists and zoologists, and will be useful to anyone looking to gain an understanding of the world's commonly encountered plant groups. Together, the two handbooks provide accounts of 149 plant families; where a family appears in both publications, the content in each book is specific to species found in the tropics or temperate zones. Along with *The Kew Plant Glossary* (2ⁿᵈ edition) (Beentje, 2016), these manuals provide a useful contribution to plant identification that is suitable for a broad audience.

The Royal Botanic Gardens, Kew works globally, collaborating with numerous partner organisations to fulfil its mission to 'understand and protect plants and fungi, for the wellbeing of people and the future of all life on Earth'. Botanists at the Royal Botanic Gardens, Kew have a long tradition of exploring and plant collecting all over the globe, and Kew is a centre of expertise for cultivating a range of plants from around the world, especially temperate plants which are grown in the Arboretum, Alpine House and Temperate House, as well as at Wakehurst. Kew's experts have accumulated an unsurpassed practical knowledge of the plants they encounter, both scientific and horticultural.

Global temperate regions

The world's temperate regions can be defined as the areas between the subtropics and the polar circles. The climate in these regions is varied and usually has distinct seasons, although the nature and duration of these seasons depend on latitude. The temperate zones of the globe comprise many different vegetation types, such as Mediterranean biomes, deciduous forests, temperate rainforests, coniferous forests and meadows and prairies, each rich in plant life. The 100 plant families that are included in this handbook have been selected because they are commonly encountered in temperate regions. Some families, for example Compositae, Iridaceae and Poaceae, are species-rich across most temperate biomes. Other families, such as Proteaceae (southern hemisphere) and conifers (northern temperate), are species-rich in particular areas. Several of the featured families, such as Equisetaceae or Dennstaedtiaceae, do not contain huge numbers of species but are represented by a few very common genera or species. Some of the included families, such as Lamiaceae, have a cosmopolitan distribution; in these cases, the family descriptions have been modified so that they represent taxa found in temperate areas rather than the family across its entire distribution. For Leguminosae and Malvaceae, families that have significant global diversity, only the subfamilies that are most species-rich in temperate areas are presented. Given the nature of plant distributions, there will always be a degree of overlap in descriptive data; occasionally details that are applicable to taxa from subtropical areas are included for completeness. Similarly, the images have been selected to demonstrate key characters for each plant family; where possible, these images are from taxa that have temperate distributions. Occasionally, images that are subtropical in origin are used if they are deemed to illustrate a family character more effectively than the available images taken in temperate zones, or if a species or genus is conservative in its features but has a distribution that spans temperate and subtropical regions.

Plant family delimitation

The sequence of flowering plant families (Aristolochiaceae to Apiaceae) and the circumscription of families follows the latest classification adopted by the botanical community, based largely on molecular findings and outlined in the papers of the Angiosperm Phylogeny Group, currently APG IV (2016). There are 416 plant families in the APG IV system (see Stevens (2001 onwards) for family information). Many Floras, however, were begun before the recent APG classifications and, thus, some plant family names that exist in the floristic literature are circumscribed differently in the new system. For example, treatments of the Scrophulariaceae have been published within several Floras that cover temperate regions (see Webb's (1972) *Flora Europaea* account, for example), but this family has now been restructured to give an altered Scrophulariaceae in its strict sense, with genera that were formerly in the Scrophulariaceae now transferred to Plantaginaceae, Orobanchaceae (both covered in this book) and several other smaller plant families. Notes describing major taxonomic changes that readers should be aware of are provided where relevant. The sequence for gymnosperms follows Christenhusz *et al.* (2011). For the ferns, we have been guided by Christenhusz & Chase (2014) and Christenhusz, Fay & Chase (2017) in *Plants of the World, an Illustrated Encylopedia of Vascular Plants*, which takes a broad view of fern families, particularly in the order Polypodiales.

How to use this book

Each family has a double-page spread, the first page with easily recognisable characters and family descriptions, herbarium specimen images and line drawings, and the second with photographs of the plants, either from the field or in cultivation. Terminology follows *The Kew Plant Glossary* (Beentje, 2016).

The value of the book in helping users to identify a plant is not conveyed in the descriptions alone. We have attempted to provide several 'layers' of information to allow the user multiple ways of identifying a plant. We hope that a broad range of users will be able to find the information accessible. Thus, in addition to the main, formal description, we provide a set of general key characters to help comparisons with other families, then a short description that highlights the

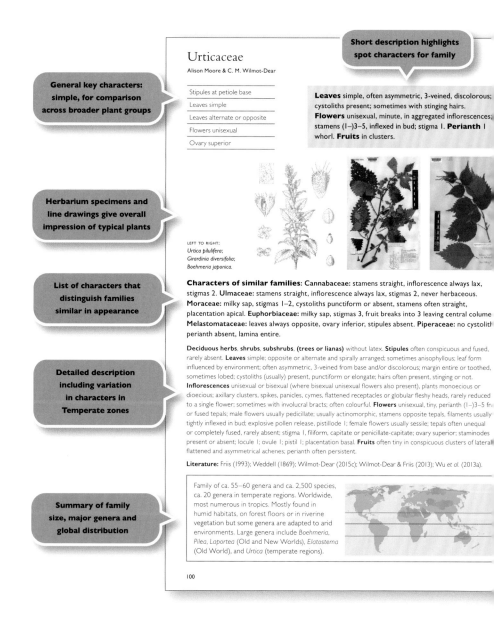

General key characters: simple, for comparison across broader plant groups

Short description highlights spot characters for family

Herbarium specimens and line drawings give overall impression of typical plants

List of characters that distinguish families similar in appearance

Detailed description including variation in characters in Temperate zones

Summary of family size, major genera and global distribution

Urticaceae
Alison Moore & C. M. Wilmot-Dear

Stipules at petiole base

Leaves simple

Leaves alternate or opposite

Flowers unisexual

Ovary superior

Leaves simple, often asymmetric, 3-veined, discolorous; cystoliths present; sometimes with stinging hairs.
Flowers unisexual, minute, in aggregated inflorescences; stamens (1–)3–5, inflexed in bud; stigma 1. **Perianth** 1 whorl. **Fruits** in clusters.

LEFT TO RIGHT:
Urtica pilulifera;
Girardinia diversifolia;
Boehmeria japonica.

Characters of similar families: Cannabaceae: stamens straight, inflorescence always lax, stigmas 2. **Ulmaceae:** stamens straight, inflorescence always lax, stigmas 2, never herbaceous. **Moraceae:** milky sap, stigmas 1–2, cystoliths punctiform or absent, stamens often straight, placentation apical. **Euphorbiaceae:** milky sap, stigmas 3, fruit breaks into 3 leaving central columne **Melastomataceae:** leaves always opposite, ovary inferior, stipules absent. **Piperaceae:** no cystolith perianth absent, lamina entire.

Deciduous herbs, shrubs, subshrubs, (trees or lianas) without latex. **Stipules** often conspicuous and fused, rarely absent. **Leaves** simple; opposite or alternate and spirally arranged; sometimes anisophyllous; leaf form influenced by environment; often asymmetric, 3-veined from base and/or discolorous; margin entire or toothed, sometimes lobed; cystoliths (usually) present, punctiform or elongate; hairs often present, stinging or not. **Inflorescences** unisexual or bisexual (where bisexual unisexual flowers also present), plants monoecious or dioecious; axillary clusters, spikes, panicles, cymes, flattened receptacles or globular fleshy heads, rarely reduced to a single flower; sometimes with involucral bracts; often colourful. **Flowers** unisexual, tiny, perianth (1–)3–5 fr or fused tepals; male flowers usually pedicillate; usually actinomorphic, stamens opposite tepals, filaments usually tightly inflexed in bud; explosive pollen release, pistillode 1; female flowers usually sessile; tepals often unequal or completely fused, rarely absent; stigma 1, filiform, capitate or penicillate-capitate; ovary superior; staminodes present or absent; locule 1; ovule 1; pistil 1; placentation basal. **Fruits** often tiny in conspicuous clusters of lateral flattened and asymmetrical achenes; perianth often persistent.

Literature: Friis (1993); Weddell (1869); Wilmot-Dear (2015c); Wilmot-Dear & Friis (2013); Wu et al. (2013a).

Family of ca. 55–60 genera and ca. 2,500 species, ca. 20 genera in temperate regions. Worldwide, most numerous in tropics. Mostly found in humid habitats, on forest floors or in riverine vegetation but some genera are adapted to arid environments. Large genera include *Boehmeria*, *Pilea*, *Laportea* (Old and New Worlds), *Elatostema* (Old World), and *Urtica* (temperate regions).

100

spot characters most useful for the family. The general key characters have been chosen from lists standardised for each of four groups: ferns, gymnosperms, monocotyledonous plants and dicotyledonous plants; users should be aware that although these key characters represent the typical character states for each family, exceptions are likely. Spot characters tend to be features that are easily seen by eye, or with a ×10 hand lens, so they can be used in the field, where the detailed study of tiny features using microscopes or dissections is not possible. Unfortunately, plant diversity has left us with no way to avoid the numerous 'often', 'usually' and 'sometimes' in front of many characters in these descriptions! Each section is explained further in a mock-up of a family treatment, shown below.

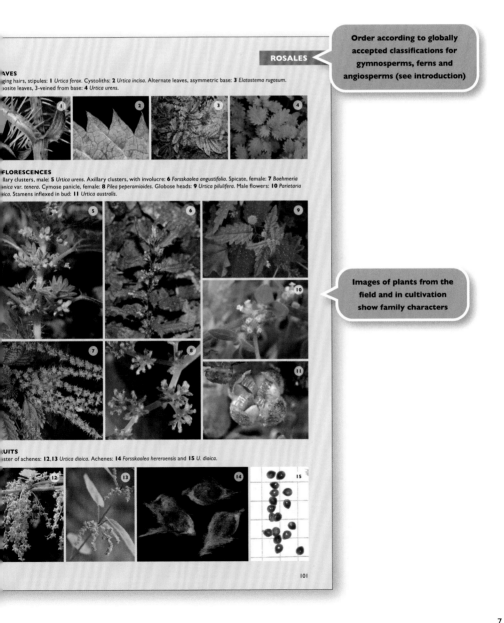

ROSALES

Order according to globally accepted classifications for gymnosperms, ferns and angiosperms (see introduction)

AVES
ging hairs, stipules: **1** *Urtica ferox*. Cystoliths: **2** *Urtica incisa*. Alternate leaves, asymmetric base: **3** *Elatostema rugosum*.
osite leaves, 3-veined from base: **4** *Urtica urens*.

FLORESCENCES
llary clusters, male: **5** *Urtica urens*. Axillary clusters, with involucre: **6** *Forsskaolea angustifolia*. Spicate, female: **7** *Boehmeria*
nica var. tenera. Cymose panicle, female: **8** *Pilea peperomioides*. Globose heads: **9** *Urtica pilulifera*. Male flowers: **10** *Parietaria*
ica. Stamens inflexed in bud: **11** *Urtica australis*.

Images of plants from the field and in cultivation show family characters

UITS
ster of achenes: **12,13** *Urtica dioica*. Achenes: **14** *Forsskaolea hereroensis* and **15** *U. dioica*.

101

7

Equisetaceae

Aurélie Grall

Terrestrial or aquatic
Rhizome present
Stem hollow with nodes
Leaves microphyllous
Spores borne on strobili

Terrestrial or aquatic **ferns**. **Stem** ribbed, hollow, and with a typical jointed appearance. **Leaves** microphyllous, small, scale-like, whorled and fused into nodal sheaths. **Sporangia** borne in strobili.

LEFT TO RIGHT:
Equisetum arvense;
Equisetum arvense x *pratense*.

Characters of similar families: Plantaginaceae (*Hippuris*): small flowers in the axils of the upper whorls of leaves, unridged stem. **Lycopodiaceae**: microphylls not organised in whorls. **Haloragaceae** (*Myriophyllum*): small green flowers in the axils of the whorls of leaves, unridged stem, pinnately divided leaves.

Small to medium-sized aquatic or terrestrial perennial **herbs**. **Aerial stems** annual or perennial, erect, monomorphic or dimorphic (e.g. *Equisetum arvense*), cylindrical, hollow, and ribbed; if branching occurs, the branches stem from the nodes. **Underground stem or rhizome** often very deep and creeping but also ascending, sometimes bearing tubers, branched, with nodes producing adventitious roots. **Leaves** single-veined (microphyllous), small, simple, scale-like, usually toothed, arranged in a whorl and united around the stem, forming a crown-like sheath at the nodes. **Strobili** cone-like, ellipsoid or terete, stalked or sessile, terminal on stems or branches, and consisting of numerous peltate umbrella-shaped sporangiophores bearing several sporangia. **Spores** homosporous, spherical, green (chlorophyllous), bearing four hygroscopic elaters.

Literature: Christenhusz *et al.* (2019); Hauke (1990); Plants of the World Online (2019); Pryer *et al.* (2001); Zhang & Turland (2013).

Equisetaceae were long considered to be 'fern allies', but modern classifications now place this family in the true ferns or monilophytes (Pryer *et al.* 2001). The most recent treatment accepts 18 species in *Equisetum* (Christenhusz *et al.* 2019). Found throughout the world from S America and Africa to above the Arctic circle, although the greatest diversity is concentrated in the northern hemisphere.

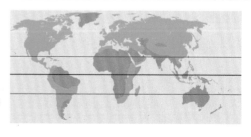

HABIT

Terrestrial: **1** *Equisetum sylvaticum*. Aquatic: **2** *Equisetum fluviatile*.

STEMS

Tortuous stems: **3** *Equisetum scirpoides*. Young stem without developed branches: **4** *Equisetum braunii*. Ribbed stem, nodal branching and nodal sheath: **5** *Equisetum telmateia*. Young plant with rhizome and adventitious roots: **6** *Equisetum* x *meridionale*.

STROBILI

Non-photosynthetic fertile stems bearing terete strobili with umbrella-shaped sporangiophores: **7, 8** *Equisetum arvense*. Ellipsoid strobili on lateral branching: **9** *Equisetum laevigatum*.

Cyatheaceae

Marcus Lehnert

Rhizomes massive, erect
Scales and hairs present
Sori round/globose
Indusia absent or present
Indusia partial to entire

Ferns with **rhizomes** radial, mostly erect, massive, forming a trunk with a distinct pattern of stipe scars. **Fronds** mostly large, decompound, basally truncate to tapering, sori spherical, dorsally on veins, with indusia or naked.

LEFT TO RIGHT:
Alsophila smithii;
Gymnosphaera podophylla;
Sphaeropteris cooperi.

Characters of similar families: **Dicksoniaceae** and **Cibotiaceae**: massive rhizomes creeping to erect, indumentum only hairs, sori elliptic, marginal, indusia bivalved. **Saccolomataceae**: rhizomes ascending, often bent, adventitious roots loose, fleshy, indumentum mostly scales, blades triangular with asymmetric lower pinnae, sori submarginal to marginal, indusia pouch-like. **Blechnaceae** (Aspleniaceae–Blechnoideae): caudices formed mainly by stipe bases and roots, fronds pinnatifid to pinnate, segments entire, sori linear, in two lines along midrib of each segment, indusia open towards midrib.

Perennial herbs terrestrial and saxicolous, sometimes climbers and epiphytes. Some species are good indicators for humidity, soil fertility, and biotope disturbance. **Rhizomes** short-creeping to erect, trunk-like, adventitious roots fleshy to fibrous, dense, larger plants with supporting buttresses, petiole bases persisting or shedding, revealing a characteristic pattern of scars. **Mucus** flowing from injuries clear, colourless to orange, odourless, antiseptic. **Fronds** spirally arranged in dense pseudo-whorls, rarely remote, monomorphic, rarely hemidimorphic. **Petioles** inermous to aculeate, loosely to densely covered in scales of various sizes and colours, often with a dense layer of minute scales ('scurf'), persisting or ephemeral. **Blades** entire to tripinnate-pinnatifid, mostly bipinnate-pinnatifid (temperate taxa), with a species-specific fine indumentum of small scales and hairs, sometimes lamina glaucous below. **Sori** on the back or in the fork of a vein, proximal to marginal, spherical to weakly conical. **Indusia** absent or present, scale-like, discs or cups of various depths, to closed globes with apical protrusion ('umbo'), characteristically fragmenting at maturity, sometimes substituted by spirally arranged scales or a mucose layer.

Literature: Holttum (1964); Korall *et al.* (2007); Large & Braggins (2009).

Pantropical and southern temperate, with >685 taxa in four genera and several distinct subclades. From humid lowland forests to above the tree line, most diverse in the mountains at 1000–2500 m. Main diversity in the Neotropics (ca. 335 spp.) and Australasia (ca. 290 spp.); continental Africa with just 12 species.

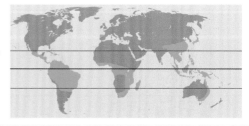

HABIT
Trunk shedding old petioles: **1** *Sphaeropteris medullaris*. Trunk retaining old petioles: **2** *Alsophila spinulosa*. Shaded plant with fronds held upright, showing glaucous underside: **3** *Alsophila tricolor* (syn. *Cyathea dealbata*). Plant with skirt of dead fronds: **4** *Alsophila smithii*.

FRONDS AND PETIOLES
Plant with ovate elliptic fronds: **5** *Alsophila colensoi*. Crozier with medium brown scales and green petioles: **6** *Alsophila kermadecensis*. Dark scales of trunks and petioles, contrasting with the glaucous layer covering the shiny black epidermis: **7** *Alsophila ferdinandii* (syn. *Cyathea macarthurii*).

SORI AND INDUSIA
Glaucous lower side with dark sori: **8** *Alsophila tricolor*. Sori position medial, lacking indusia: **9** *Alsophila colensoi*. Sori close to midvein, indusia semi-circular: **10** *Alsophila smithii*. Empty spherical indusia fragmented: **11** *Sphaeropteris medullaris*.

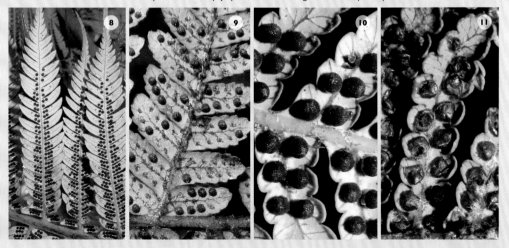

Dennstaedtiaceae

Michael Sundue

Rhizomes creeping
Hairs present
Leaves decompound
Sori marginal
Indusia present

Medium to large **ferns** with long-creeping rhizomes. **Petioles** bearing epipetiolar buds that produce new rhizomes. **Sori** marginal, discrete or continuous. Abaxial and adaxial **indusia** usually present.

LEFT TO RIGHT:
Dennstaedtia wilfordii type specimen; *Dennstaedtia glauca*.

Characters of similar families: Pteridaceae: differ in having no true indusia. **Lindsaeaceae** and **Saccolomataceae**: previously treated in **Dennstaedtiaceae** and difficult to distinguish from it, but they lack epipetiolar buds and often have short-creeping rhizomes; they are strictly tropical. **Dicksoniaceae**: some tree ferns, particularly those that lack trunks (such as *Calochlaena* and *Thyrsopteris*) could be confused with Dennstaedtiaceae, but they have sporangia with an oblique and continuous annulus.

Medium to large herbaceous perennial **ferns**, often of open and disturbed habitats. **Rhizomes** solenostelic, sometimes polycyclic; long-creeping, rarely short-creeping; pubescent, sometimes scaly. **Petioles** with omega-shaped vascular bundles, the petiole base often bearing epipetiolar buds that produce new rhizomes (occasionally petioles with nectaries). **Leaves** decompound, sometimes indeterminate and then very large and scandent, occasionally with prickles. **Veins** free, sometimes anastomosing. **Sori** marginal, sometimes submarginal; discrete or continuous. **Indusia** present, rarely absent; in various combinations of a recurved leaf margin and an abaxial indusium that together enclose the sorus, often forming a cup-like receptacle in species with discrete sori. **Sporangia** long-stalked, thin-walled, with a vertical interrupted annulus. **Spores** monolete or trilete.

Literature: Schwartsburd *et al.* (2020); Shang *et al.* (2018); Wolf *et al.* (2019).

A cosmopolitan family that is most diverse in the tropics, includes 10 genera and about 265 species. Unlike most ferns, Dennstaedtiaceae often occupy open and disturbed habitats. Temperate genera include *Dennstaedtia* and *Pteridium*, the latter includes bracken and is of major economic importance as a noxious weed. Circumscription of the genus *Dennstaedtia* is ongoing.

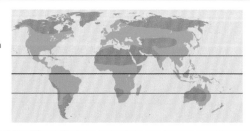

HABIT

Herbaceous perennial ferns, often of open and disturbed habitats: **1** *Pteridium arachnoideum*, **2** *Pteridium aquilinum*, and **3** *Dennstaedtia punctilobula* in autumn.

LEAVES

Leaves decompound: **4** *Pteridium latiusculum*. Leaves 2-pinnate-pinnatifid: **5** *Dennstaedtia punctilobula*. Leaves unfurling: **6** *Pteridium aquilinum*.

UPPER LEAF SURFACES AND SORI

Upper leaf surface: **7** *Pteridium arachnoideum*. Lower leaf surface: **8** *P. arachnoideum*. Sori marginal, continuous: **9** *Pteridium latiusculum*.

13

Pteridaceae

Aurelie Grall

Rhizome present
Scales and hairs present
Fronds divided
Sori marginal or along veins
False indusium often present

Small to large **ferns** with creeping to erect rhizomes. **Sporangia** in patches or fused in coenosori that are located along the veins centrally or marginally on the ventral side of the frond. True indusia missing, with the sori often protected by the in-rolled lamina margin instead (**pseudo-indusium**).

LEFT TO RIGHT:
Hemionitis michelii (syn. *Cheilanthes argentea*); *Cryptogramma acrostichoides*: fertile and sterile dimorphic fronds.

Characters of similar families: **Dennstaedtiaceae**: true indusium present. **Dryopteridaceae** (Polypodiaceae–Dryopteridoideae): sori usually with an orbiculate to reniform indusium. **Aspleniaceae** (Aspleniacae–Asplenioideae): true indusium present (an elongated flap running along veins); scales always clathrate and often iridescent. **Athyriaceae** (Aspleniacae–Athyrioideae): true indusium present; mostly terrestrial ferns growing in the understorey below trees and shrubs. **Polypodiaceae**: scales often peltate, leaves articulated to a persistent stipe base.

Small to large terrestrial and lithophytic **ferns**. **Rhizomes** creeping, ascending or erect, and in most cases scaly. **Fronds** mostly monomorphic, occasionally dimorphic. **Lamina** pinnatifid to 6 times pinnate, more rarely peltate (e.g. *Adiantum pedatum*); petiole with usually 1 or 2 vascular bundles, glabrous, hairy or scaly, glossy brown to black or duller, grey to green; veins free or anastomosing, never with included veinlets. Whole fronds sometimes able to curl up entirely in the dry season (subfamily Cheilanthoideae). Ultimate segments sometimes very reduced, often petiolate, and occasionally with a crosswise groove or articulation line. Abaxial side of the lamina sometimes covered with a white, silver or golden powder or farina or with many hairs or scales. **Sori** in many species are fused into coenosori that follow either the veins in the middle of the lamina or the lamina margin, continuously or discontinuously. The sori can also form smaller patches on vein tips (e.g. *Anogramma*) or at the bottom of the sinuses between two lobes; when marginal, the sori are often protected by a recurved portion of the lamina known as false indusium (e.g. *Adiantum, Pteris*). **Sporangia** on a long stalk. Spores mostly brown, yellowish, or colourless, mostly tetrahedral-globose and trilete, rarely ellipsoid and monolete.

Literature: Christenhusz *et al.* (2017, 2018); Hevly (1963); Schuettpelz *et al.* (2007); Tryon (1990).

Found from mountains to tropical forests. A large, diverse family with 5 subfamilies (Parkerioideae, Cryptogrammoideae, Pteridioideae, Vittarioideae and Cheilanthoideae), 53 genera and over 1000 species recognised by the Pteridophyte Phylogeny Group in 2016. Christenhusz *et al.* (2017) listed only 45 genera before lumping all cheilanthoid ferns into *Hemionitis* (Christenhusz *et al.* 2018), leaving 32 genera.

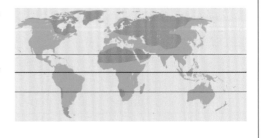

HABIT AND FRONDS

Lithophytic on a humid limestone boulder, bipinnate: **1** *Adiantum capillus-veneris*. Terrestrial in humid soil, pedate venation: **2** *Adiantum pedatum*. Lithophytic on dry rocky outcrops, pinnate: **3** *Hemionitis marantae*. Xerophytic habit with enrolment of fronds in the dry season and powdery indument (farina) on underside of frond: **4** *Hemionitis michelii*. **5**. *Hemionitis* sp.

SORI

Marginal continuous (**6** *Pteris multifida*) and interrupted (**7** *Adiantum capillus-junonis*) coenosori with pseudoindusia. Naked sori in patches under the lamina segments: **8** *Anogramma leptophylla*.

Aspleniaceae

Richard Wilford

Rhizome with wiry roots
Scales and hairs present
Leaves simple or divided
Sori along veins
Indusia usually present

Terrestrial, epiphytic or lithophytic **ferns**. **Rhizomes** creeping, erect or ascending, with wiry roots. **Petioles** often scaly at base, with 2 vascular bundles. **Leaves** simple or pinnatifid to 4-pinnate. **Sori** borne on veins.

LEFT TO RIGHT:
Gymnocarpium oyamense;
Asplenium australasicum;
Asplenium adnatum type specimen.

Characters of similar families: Pteridaceae: true indusia missing, sori often protected by the in-rolled lamina margin instead (false indusium). **Polypodiaceae:** rhizomes often fleshy, petiole with usually 3 or more vascular bundles organised to form a U-shape or semi-circle in cross-section.

Terrestrial, epiphytic or lithophytic **ferns**, rarely climbing; herbaceous or evergreen; with blackish, wiry roots. **Rhizomes** usually fibrous, creeping, erect or ascending, scaly; scales lanceolate to linear, often clathrate, translucent or opaque. **Petioles** often scaly at base; with 2 vascular bundles back-to-back and lunate to round or oblong, or more than 2 vascular bundles arranged in an arc, sometimes 1 vascular bundle X-shaped in cross-section. **Leaves** often monomorphic but sometimes strongly dimorphic with fertile leaves shorter or taller and more erect than sterile ones; sometimes reddish when young (e.g. Blechnoideae); usually clustered and spirally arranged, sometimes remote, papery to leathery; linear to lanceolate, or ovate to elliptic or oblanceolate; simple to pinnatifid to pinnate-pinnatifid to 4-pinnate, commonly with tiny glandular hairs or linear scales, rarely with spreading hairs; rachis frequently grooved adaxially; veins free, simple or forked, sometimes anastomosing. **Sori** usually along a vein and linear, or in a vein fork, rarely circular and atop a vein, occasionally nearly marginal, often fused into coenosori between midrib and margin, rarely acrostichoid; indusia usually present and persistent, often laterally attached, sometimes cup-shaped, rarely absent. **Sporangia** stalked; spores mostly brownish.

Literature: Christenhusz & Chase (2014); Christenhusz *et al.* (2017); Flora of North America Editorial Committee (1993); Plants of the World Online (2021); Wu *et al.* (2013b).

Distributed worldwide with the exception of frozen and arid areas. A family of 24 genera (ca. 16 in temperate regions) and ca. 2,800 species if families such as Blechnaceae, Athyriaceae, Thelypteridaceae and Woodsiaceae are included as subfamilies within Aspleniaceae (Christenhusz & Chase, 2014). Temperate species, including species of *Athyrium*, *Asplenium* and *Blechnum*, are often used as garden ornamentals.

HABIT

Small lithophytic fern: **1** *Asplenium trichomanes* growing on a stone wall. Terrestrial, evergreen, simple leaves: **2** *Asplenium scolopendrium*. Creeping rhizomatous perennial: **3** *Blechnum penna-marina*.

LEAVES

Young leaves reddish: **4** *Woodwardia unigemmata*. Bipinnate: **5** *Athyrium niponicum*. Leathery, pinnate: **6** *Blechnum chilense*. Dimorphic fronds with fertile fronds pinnate and shorter than sterile: **7** *Onoclea struthiopteris*.

SORI

Linear coensori along veins: **8** *Asplenium scolopendrium*. Round, on veins of pinnae between midrib and margin: **9** *Cystopteris fragilis*. Round with cup-shaped indusia divided into shallow lobes: **10** *Woodsia polystichoides*.

Polypodiaceae

Richard Wilford

Rhizome often fleshy
Scales and hairs present
Leaves simple or divided
Sori borne on veins
Indusia present or absent

Terrestrial, lithophytic or epiphytic **ferns.** Evergreen or deciduous. **Rhizome** creeping or erect, often fleshy. **Petiole** with usually 3 vascular bundles, sometimes more. **Leaves** simple, pinnatifid to 4-pinnate. **Sori** borne on veins.

LEFT TO RIGHT:
Polypodium vulgare;
Lepisorus contortus.

Characters of similar families: Pteridaceae: petiole with usually 1 or 2 vascular bundles, true indusia missing, sori often protected by the in-rolled lamina margin instead (false indusium). Aspleniaceae: rhizomes with wiry roots, rarely fleshy, vascular bundles usually 2 or sometimes forming an X shape, if more than 2 vascular bundles then not in a U-shape.

Terrestrial (many temperate genera), lithophytic or epiphytic **ferns**, evergreen or deciduous. **Rhizome** long- to short-creeping or erect, often fleshy, dictyostelic; bearing scales; scales clathrate or opaque, basally attached or peltate; roots fleshy or wiry. **Petiole** usually articulate at base (not articulate in subfamily Dryopteridoideae); usually with 3, sometimes more, vascular bundles, arranged in a semi-circle or U shape in cross-section. **Leaves** monomorphic or dimorphic, caespitose or remote, sometimes spirally arranged, circinate in bud; broadly ovate to lanceolate, linear or deltate, mostly simple to pinnatifid, pinnatisect, or up to 4-pinnate; thinly papery, papery, or leathery. Veins free, simple, forked or pinnate or often anastomosing; indument various, of scales, hairs, or glands. **Sori** borne abaxially on leaf veins or ends of veins, superficial or impressed, round to oblong, occasionally elongate, rarely marginal or acrostichoid and covering abaxial surface; indusia absent or variously linear, falcate, or reniform, sometimes hood-like, cup-like, or round (e.g. subfamily Dryopteridoideae). **Sporangia** stalked; spores usually transparent or yellowish to brownish, rarely greenish or black.

Literature: Christenhusz & Chase (2014); Christenhusz *et al.* (2017); Flora of North America Editorial Committee (1993); Plants of the World Online (2021); Wu *et al.* (2013b).

Found worldwide but absent from frozen and arid areas. With the inclusion of genera previously in Dryopteridaceae, this is the largest family of ferns, with ca. 70 genera and over 4,000 species. Many are tropical or subtropical but large temperate genera include *Dryopteris*, *Lepisorus*, *Polystichum* and *Polypodium*. Temperate species are often used as garden ornamentals.

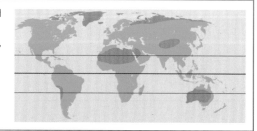

HABIT

Terrestrial, creeping evergreen perennial: **1** *Polypodium vulgare* and **2** *Pyrrosia lingua*. Rhizome erect, clump-forming: **3** *Dryopteris wallichiana*. Lithophytic with creeping rhizome: **4** *Davallia trichomanoides*.

LEAVES

Fronds pinnate: **5** *Dryopteris cycadina*. Imparipinnate: **6** *Dryopteris sieboldii*. Bipinnate: **7** *Polystichum setiferum*.

SORI

Round, indusia absent: **8** *Polypodium vulgare*. Arranged between main lateral veins, indusia absent: **9** *Pyrrosia sheareri*. In pairs on each segment, indusia present: **10** *Dryopteris wallichiana*. In irregular rows, indusia present: **11** *Dryopteris sieboldii*.

19

Cycadaceae and Zamiaceae

Richard Wilford

Stem aerial or subterranean
Leaves compound
Plants dioecious
Pollen cones terminal
Seed held on megasporophylls

Trees or perennial **herbs**, dioecious, with short stem or trunk, sometimes tall and palm-like. **Leaves** coriaceous, evergreen, mostly pinnate. **Pollen cones** terminal. **Seeds** on aggregated megasporophylls (Cycadaceae) or seed cones (Zamiaceae).

LEFT TO RIGHT:
Fruiting *Cycas circinalis*;
Stangeria eriopus cones.

Characters of similar families: Palmae: compound leaves derived by splitting of an initially entire lamina, flowers in lateral inflorescences, fruit a berry or drupe.

Trees or perennial **herbs**, dioecious, with a subterranean or aerial stem or trunk clothed with persistent leaf bases; subterranean stems have their apex at ground level; aerial stems often do not exhibit axillary branching; some adventitious roots branched and shaped like coral, and containing symbiotic, nitrogen-fixing cyanobacteria. **Leaves** spirally clustered at stem apex, petiolate, pinnate (bipinnate in *Bowenia*–Zamiaceae), coriaceous, evergreen, leaflets entire or dentate, sometimes with spines; leaflets or leaves sometimes exhibiting circinate vernation (*Cycas*, *Stangeria*). **Pollen cones** (strobili) are produced at the apex of the stem, consisting of microsporophylls bearing numerous microsporangia; sperm is free-swimming. **Female megasporophylls** arranged in a cone (Zamiaceae) or free but aggregated in a dense mass (Cycadaceae), at the apex of the stem. **Seeds** have a hard inner and fleshy outer seed coat, large and often brightly coloured.

Literature: Christenhusz *et al.* (2017); Simpson (2019).

Cycadales comprises 2 families: Cycadaceae with 1 genus (*Cycas*) of ca.100 species, and Zamiaceae with 9 genera and ca. 200 species. Distribution mostly tropical or subtropical but extending into temperate regions, such as Japan, South Africa and Australia. *Cycas* is a source of starch (sago) but poisonous if untreated. Some species, such as *Cycas revoluta*, are used ornamentally.

HABIT

General habit: **1** Cycad grove at Kirstenbosch Botanic Garden, South Africa. Trunk unbranched: **2** *Encephalartos altensteinii* in the Palm House at Kew. **3** *Cycas revoluta* planted as an ornamental.

LEAVES

Pinnate leaves: **4** *Encephalartos transvenosus*. Spirally clustered at stem apex: **5** *Encephalartos woodii*. Leaflets with spines: **6** *Encephalartos horridus*.

CONES

Pollen cone(s) at apex of stem: **7** *Ceratozamia kuesteriana*, **8** *Encephalartos altensteinii* and **9** *Dioon spinulosum*.

Pinaceae

Harry Baldwin & Tony Kirkham

Leaves acicular-linear
Leaves whorls or fascicles
Pollen cones solitary or clustered
Seed scales and bracts free
Two seeds per scale

Trees or **shrubs**, evergreen or deciduous. **Leaves** usually aromatic when crushed, acicular, often spirally arranged in fascicles. **Seed cones** woody, scales spirally arranged, scales and bracts free, seeds usually two per scale.

LEFT TO RIGHT:
Picea jezoensis subsp. *jezoensis*;
Larix griffithii.

Characters of similar families: Araucariaceae: seed cones consisting of bracts forming the cone scales only; seeds 1–many. **Sciadopitiaceae:** true leaves reduced to small cataphylls (turning to brown scales); cladodes in pseudo-whorls. **Taxaceae:** seed cones reduced to a single seed, partly or completely surrounded by an aril, seed terminally placed on a scaly dwarf shoot. **Podocarpaceae:** true leaves scale-like, acicula or with distinct lamina; cones arising in axils of leaves. **Cupressaceae:** leaves scale-like or acicular; seed cones, bracts and scales fused with multiple seeds per scale.

Trees or shrubs, monoecious, evergreen or deciduous, aromatic and resinous. **Branching** in rhythmic pseudo-whorls on main stem and branches. **Shoots** glabrous, pubescent or hairy and resinous buds in dormant season. **Leaves** usually evergreen (deciduous in *Larix* and *Pseudolarix*), acicular, linear to long, in whorls (*Larix* and *Pseudolarix*) or spirally arranged in fascicles of 1–8 surrounded by a sheath on immature shoots (*Pinus*), on short shoots in some taxa (*Cedrus, Larix* or *Pseudolarix*). **Pollen cones** often grouped close together on long shoots, axillary, solitary or clustered from a single bud, catkin-like, deciduous; microsporophylls numerous, spirally arranged with 2 abaxial pollen sacs. **Seed cones** woody, small to large, erect or pendulous, lateral on long shoots or apical on short shoots, solitary in leaf axils. Scales spirally arranged on a woody rachis (breaking up in *Pseudolarix*), remaining small or growing with the cone; bracts apical often exserted (*Pseudotsuga*); ovules usually 2, adaxial, subtending bracts free from ovuliferous scale. **Seeds** usually 2 per scale; scales free from bract, persistent or deciduous (*Abies* or *Cedrus*); seeds usually winged or unwinged.

Literature: Bittrich *et al.* (1993); Chase *et al.* (2017); Debreczy & Racz (2011).

Pinaceae consists of eleven genera that are distributed almost solely throughout the Northern Hemisphere, with only *Pinus merkusii* entering the Southern Hemisphere. As the largest conifer family, Pinaceae consists of ca. 240 species, many of which have great ecological and economic importance.

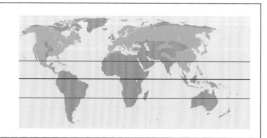

HABIT

Large evergreen tree: **1** *Pinus nigra* and **2** *Tsuga mertensiana*. Deciduous tree: **3** *Larix decidua*.

LEAVES

Needles spirally arranged: **4** *Cedrus atlantica* 'Glauca'. Leaves spirally arranged on branchlets in dense clusters: **5** *Larix gmelinii* var. *olgensis*. Emerging needles: **6** *Picea likiangensis*.

CONES

Pollen cone: **7** *Larix kaempferi*. Female cones and new growth: **8** *Pinus thunbergii*. Seed cone: **9** *Abies koreana*. Mature cone: **10** *Picea likiangensis*. Seed scales terminating in sharp spines: **11** *Pinus pungens*.

23

Araucariaceae

Martin Xanthos

Leaves helically arranged

Lamina broad and flat

Pollen cones solitary or clustered

Seed cone scales fused with bracts

Single inverted seed per scale

Highly resinous **trees. Leaves** helically attached, lamina broad and flat or scale-like. **Pollen cones** catkin-like, sometimes large. **Seed cones** large, globose. Seed cone scales fused with the bract enclosing a single inverted seed.

LEFT TO RIGHT:
Araucaria angustifolia;
Araucaria bidwillii;
Agathis microstachya.

Characters of similar families: Pinaceae: seeds 2, on the upper side of each fertile scale, seed cone scales in the axils of bracts. **Cupressaceae:** seeds 1–many, axillary or on the base of each bract.

Dioecious or monoecious evergreen, highly resinous **trees**, bark rough and exfoliating in horizontal strips. Branches in pseudowhorls, spreading or ascending. **Foliage branchlets** with or without terminal buds. Leaves spirally arranged (opposite to subopposite in *Wollemia*), laminae flat and triangular to lanceolate or scale-like, sessile, imbricately covering the shoot and more or less distichously spreading (4-ranked in *Wollemia*), more or less coriaceous. **Pollen cones** axillary to the leaves, solitary or in small clusters, much elongating after anthesis and becoming cylindrical. Microphylls numerous, helically inserted, crowded with imbricate or tessellate heads, each with 4–20 oblong pollen sacs. **Seed cones** terminal on long shoots or lateral on short, pedunculate, leafy shoots, solitary, erect, ovoid or subglobose, usually disintegrating leaving its rachis on the tree. Bracts helically inserted on the rachis, much developed, flattened, with a thickened distal margin and with or without a terminal elongated cusp, forming the bulk of the cone. Seed scales much reduced, axillary to and almost entirely fused with the bract with or without a small free apical ligule enclosing the seed. **Seeds** wingless or with 1 or 2 unequal wings.

Literature: Farjon (2017).

Three genera with 37 species. Occurs in all major islands in Malesia except Java and the lesser Sunda Islands, in Australia (New South Wales and Queensland), on the SW Pacific Islands, in New Zealand (North Island), and in South America (SE Brazil, NE Argentina, S Chile, and SW Argentina (Andes)). Temperate species of *Araucaria* are often planted as ornamental trees.

MORPHOLOGY, LEAVES AND BARK

Branches in pseudowhorls: **1** *Agathis microstachya*. Rough exfoliating bark: **2** *Araucaria angustifolia* and **3** *Wollemia nobilis*. Spirally arranged leaves: **4** *Araucaria bidwillii* and **5** *Araucaria araucana*. Foliage in 4 ranks: **6** *Wollemia nobilis*.

POLLEN AND SEED CONES

Seed cones: **7** *Agathis australis*. Pollen cones: **8** *Araucaria angustifolia* and **9** *Araucaria araucana*.

CONE SCALES AND SEEDS

Seed cone scales: **10** *Agathis australis* and **11** *Wollemia nobilis*. Seed cone scale and seed: **12** *Araucaria araucana*.

Podocarpaceae

Harry Baldwin & Tony Kirkham

Usually dioecious
Leaves simple
Pollen cones solitary or clustered
Cones arise in leaf axils
One seed per cone

Dioecious **trees** or **shrubs**. **Leaves** usually linear, elliptic or scale-like. **Seed cones** with ovuliferous scales bearing one ovule, cone reduced to one seed, seated on fleshy receptacle, often enveloped by a fleshy false aril.

LEFT TO RIGHT:
Podocarpus macrophyllus;
Podocarpus henkelii.

Characters of similar families: Taxaceae: seed cones reduced to a single seed, partly or completely surrounded by an aril, seed terminally placed on a scaly dwarf shoot, fleshy receptacle absent. **Sciadopitiaceae:** true leaves reduced to small cataphylls (turning to brown scales); cladodes in pseudowhorls. **Cupressaceae:** leaves scale-like or acicular; seed cones, bracts and scales fused with multiple seeds per scale.

Trees or **shrubs**, evergreen, usually dioecious, rarely monoecious. **Leaves** simple, entire, often highly variable in form, from large, linear, elliptic or subulate to scale-like, often spirally arranged, rarely decussate or subopposite; shoots glabrous. **Pollen cones** terminal or axillary, solitary or clustered, often catkin-like; each with numerous spirally arranged scales each bearing 2 pollen sacs; pollen grains usually 2 sacs, sometimes 3, rarely 0 (*Saxegothaea*). **Seed cones** maturing in one year, terminal or axillary, usually solitary, varying much in complexity between genera; some subtended by a peduncle that may fuse with bracts, forming a fleshy receptacle that may be either dry or become fleshy and succulent after fertilisation, ovuliferous scales one, each bearing a single inverted ovule. **Seeds** one inverted or erect, often protruding, in some taxa with a fleshy, sometimes colourful fleshy false aril (epimatium) or dry and seated on an enlarged fleshy receptacle, functioning in bird dispersal; embryo with two cotyledons.

Literature: Bittrich *et al.* (1993); Chase *et al.* (2017); Debreczy & Racz (2011).

The second largest family of conifers, containing 18 genera and ca. 180 species. Mostly distributed in the Southern Hemisphere (often tropical, mainly Australasia to SE Asia, ranging to Japan), but also in Central and South America and tropical montane Africa. Genera that are most often cultivated in cooler regions include *Phyllocladus*, *Prumnopitys*, *Saxegothaea*, and relatively smaller portions of *Podocarpus*, *Dacrydium* and *Dacrycarpus*.

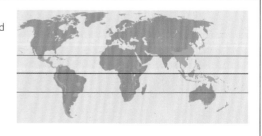

HABIT AND LEAVES

Trees: **1** *Prumnopitys andina* in Chile. Shrubs: **2** *Podocarpus acutifolius*. Prostrate shrub, twigs slender and quadrangular in cross-section: **3** *Microcachrys tetragona*. Spirally arranged sharp needles: **4** *Podocarpus laetus*.

CONES AND FRUIT

Male strobili on short, axillary shoots: **5** *Prumnopitys andina*. Seed cones terminal, turning red when ripe: **6** *Microcachrys tetragona*. Small oblong nut borne on enlarged, succulent, bright red receptacle: **7** *Podocarpus nivalis*.

Cuppressaceae

Harry Baldwin & Tony Kirkham

Leaves scale-like or acicular
Twigs often heterophyllous
Pollen cones mostly terminal
Cone scales whorls of 3 or 4
Ovules several per scale

Trees or **shrubs**. **Leaves** simple, spirally arranged, deltate to linear often appearing scale-like, twigs often heterophyllous. **Pollen cones** with 2–7 microsporangia. **Ovuliferous scales** in whorls of 3–4, each bearing 1–20 ovules.

LEFT TO RIGHT:
Juniperus sabina;
Cryptomeria japonica.

Characters of similar families: Araucariaceae: monoecious (*Agathis*, usually, and *Wollemia*), or dioecious (*Araucaria*). Leaves evergreen, often broad to acicular; pollen cones large, 5–20 microsporangia; seed cones erect and disintegrating on the tree when mature, ovuliferous scale bearing a single ovule. Pinaceae: leaves acicular, linear to long in whorls, seed cone scales free from bract, usually 2 seeds per scale. Taxaceae: pollen cones axillary, seed partially or wholly surrounded by fleshy or leathery aril.

Trees or **prostrate shrubs**, evergreen or deciduous, monoecious or dioecious, aromatic, resinous. **Bark** fibrous or brittle, exfoliating in longitudinal strips or small plates. Winter buds mostly absent. **Leaves** simple, alternate and spirally arranged, sometimes twisted appearing 2-ranked, opposite in 4 ranks, or whorled; deltate to linear, sessile or petiolate; adult leaves appressed or spreading, often twigs heterophyllous, sometimes strongly dimorphic on each twig (*Thuja*, *Thujopsis*, *Libocedrus*, and *Calocedrus*) with lateral scale-leaf pairs conspicuously keeled, glandular, crowded, green shining or slightly glaucous; juvenile leaves linear, flattened, spreading, often with a solitary abaxial resin gland. **Pollen cones** terminal, rarely axillary, solitary or sometimes clustered in groups of 2–7, sessile on foliage branches; microsporophylls spirally arranged; microsporangia abaxial, free 2–10(–14) per microsporophyll. Seed cones terminal (some appear axillary), sessile or pedunculate, solitary or rarely in clusters of 2–5 (to 100 in *Widdringtonia*); scales spirally arranged in whorls of 3–4, fused to subtending bracts, but sometimes with only the bract apex free, each scale bract complex peltate, oblong or cuneate, woody or fleshy at maturity, bearing 1–20 ovules. Seeds are wingless or with 2 or 3 wings; aril lacking; cotyledons 2–5 (to 9 in *Taxodium*).

Literature: Chase *et al.* (2017); Debreczy & Racz (2011); Farjon (2005).

Cupressaceae is on all continents except Antarctica and contains ca. 30 genera with ca. 140 species found, many of which are economically valuable. The family was formerly divided into two families: Cupressaceae, with whorled or opposite (4-ranked) leaves, and Taxodiaceae, with mostly alternate leaves. Molecular studies have shown a closer relationship, so they are now united.

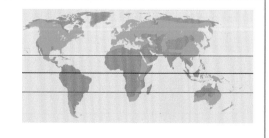

HABIT

Trees: **1** *Juniperus grandis* and **2** *Sequoia sempervirens*. Flaking bark: **3** *Cupressus bakeri*.

LEAVES

In whorls of 3, grooved with white stomatal band: **4** *Juniperus rigida*. White stomatal band abaxially: **5** *Thujopsis dolabrata*. Leaves scale-like: **6** *Thuja occidentalis*. Needles in 2 ranks: **7** *Metasequoia glyptostroboides*.

CONES

Subglobose and slightly cubic: **8** *Metasequoia glyptostroboides*. Immature female cones: **9** *Thuja occidentalis*. Pollen cones terminating on branchlets: **10** *Calocedrus decurrens*. Mature seed cone: **11** *Cryptomeria japonica*.

Taxaceae

Harry Baldwin & Tony Kirkham

Evergreen trees or shrubs
Spiral to decussate leaves
Leaves linear to acicular
Pollen cones axillary
One seed arillate

Evergreen **trees** or **shrubs**. **Leaves** linear to acicular, spiral to decussate. **Pollen cones** axillary, peltate microsporophylls each bearing 2–17 microsporangia. Mature **seed cones** reduced to 1 seed, seed arillate.

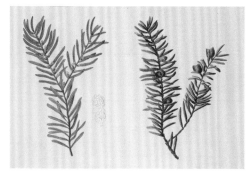

LEFT TO RIGHT:
Taxus baccata;
Cephalotaxus fortunei: holotype.

Characters of similar families: Cephalotaxaceae: there is ongoing debate over the status of this monogeneric family. The main characters used to distinguish it from Taxaceae s.s. are the aril fully enclosing the seed and the 2-ovulate bracts in the seed cones. **Cupressaceae:** pollen cones usually terminal, multiple seeds per scale, aril absent. **Pinaceae:** seed cone woody, seeds usually 2 per scale, aril absent. **Podocarpaceae:** true aril absent but seed often enveloped by a fleshy epimatium, leaves highly variable in form.

Trees or shrubs, evergreen, resinous or non-resinous (foul-odoured in *Torreya*), dioecious or monoecious. **Bark** scaly or fissured. **Leaves** simple, spirally arranged but twisted to appear in 2 ranks, linear, lanceolate to acicular, typically decurrent. **Pollen cones** solitary or clustered, axillary on year-old branches, globose to ovoid, microsporophylls peltate, each bearing 2–16 microsporangia. **Seed cones** reduced to 1 ovule (*Cephalotaxus* seed cones with decussate bracts, each subtending 2 ovules), not winged, hard seed coat partially or wholly surrounded by fleshy or leathery aril, cotyledons 2.

Literature: Majeed *et al.* (2018); Chase *et al.* (2017); Chen *et al.* (2020).

Taxaceae (including Cephalotaxaceae) has ca. 34 species in 6 genera: *Amentotaxus*, *Austrotaxus*, *Cephalotaxus*, *Pseudotaxus*, *Taxus* and *Torreya*. The family is distributed in N America, N Africa, and Eurasia to SE Asia. *Cephalotaxus*, included in the Taxaceae here, is sometimes treated as a separate, monogeneric family. *Taxus baccata* (yew) is widely grown in gardens.

HABIT

Multi-stemmed tree: **1** *Taxus cuspidata*. Trees reaching a great age: **2** *Taxus baccata*. Female specimen with mature fruiting cones: **3** *T. cuspidata*. Reddish brown, flaky bark: **4** *T. baccata*.

LEAVES

Flexible, 2-ranked, sessile: **5** *Taxus baccata*. Two-ranked, linear, apex sharply acuminate: **6** *Torreya taxifolia*.

CONES

In axils of terminal buds: **7** *Cephalotaxus fortunei*. Aril purplish red when ripe: **8** *Taxus cuspidata* 'Nana'. Yellow or green, turning purple when ripe: **9** *Cephalotaxus sinensis*. Fleshy aril becomes hardened: **10** *Torreya nucifera*.

Aristolochiaceae

Sara Edwards

Stipules absent
Leaves simple
Leaves alternate
Flowers bisexual
Ovary inferior

Climbers, **herbs** or **shrubs**, rarely trees. **Leaves** alternate. **Flowers** bisexual; perianth uniseriate, 1–3-lobed; tube short or longer, often S-shaped; stamens in 1–4 whorls in gynostemium; ovary inferior. **Fruit** usually capsular, septicidal.

LEFT TO RIGHT:
Aristolochia sempervirens;
Aristolochia bodamae;
Aristolochia lycica.

Characters of similar families: Dioscoreaceae: petiole base and apex pulvinate, flowers often unisexual, fruits usually flattened and winged. **Menispermaceae:** leaves often peltate, flowers very small, unisexual, usually 3-merous, ovary superior, fruit of several drupelets. **Piperaceae:** flowers very small arranged in spikes, perianth absent, fruits small berries or drupes.

Climbers, **herbs** or **shrubs**, rarely **trees**. **Stipules** absent, pseudostipules sometimes present. **Leaves** alternate, distichous, simple; lamina margins usually entire; venation often palmate or pinnate. **Inflorescences** solitary or in rhipidia, terminal or axillary or on old wood at base. **Flowers** bisexual, leaf-opposed, actino- or zygomorphic; perianth uniseriate (biseriate in *Saruma*), forming a short 1–3(–5)-lobed tube, or a longer often S-shaped tube with swollen base then constricted above with 1–3 expanded lobes; stamens 5–40 or more, arranged in 1–4 whorls fused with gynoecium to form gynostemium; ovary inferior, syncarpous (apocarpous in *Saruma*), 4–6 carpels, 4–6 locules; style 3–6-lobed or with many branches. **Fruit** usually capsular, septicidal from the base upwards (forming a 'hanging basket' in some *Aristolochia* spp.), or less often an indehiscent berry or schizocarp. **Seeds** numerous, usually flat, triangular.

Literature: Barringer & Whittemore (1997); Bramley & Edwards (2015); Huang *et al.* (2003); Huber (1993); Plants of the World Online (2019); Stevens (2001 onwards).

Worldwide, except the Arctic; 7–9 genera and ca. 590 species. Aristolochiaceae are used in traditional medicine in childbirth and for snake bites. The largest genus is *Aristolochia* (ca. 450 spp.). APG IV sunk *Howardia* and *Isotrema* into *Aristolochia* and Aristolochiaceae now includes the rare parasitic Hydnoraceae. Hydnoroideae is not included here as it is tropical or sub-tropical.

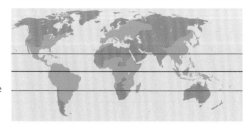

HABIT
Compact herb: **1** *Asarum maximum*. Scrambling herb: **2** *Aristolochia clematitis*. Climber: **3** *Aristolochia littoralis*.

FLOWERS
Tubular S-shaped, 3-lobed flower: **4** *Aristolochia manshuriensis*. Tubular S-shaped, 1-lobed flower: **5** *Aristolochia littoralis* and **6** *Aristolochia clematitis*. Short corolla tube, actinomorphic 3-lobed flower: **7** *Asarum* spp. and **8** *Asarum canadense*. Perianth biseriate, calyx and corolla: **9** *Saruma henryi*. Floral tube opened: **10** *Aristolochia baetica*.

FRUIT
Predehiscence: **11** *Aristolochia albida* and **12** *Saruma henryi*. Post dehiscence: **13** *A. albida*.

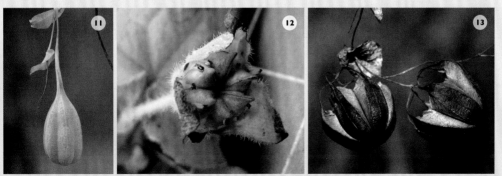

Magnoliaceae

Harry Baldwin & Tony Kirkham

Stipules present
Leaves alternate
Leaves simple
Flowers actinomorphic
Ovary superior

Trees or **shrubs**. Buds often enclosed by floccose stipules. **Leaves** simple or lobed, alternate. **Inflorescence** a solitary flower, usually hermaphrodite; tepals numerous. **Fruit** woody, dehiscent. **Seeds** red, fleshy or samaras.

LEFT TO RIGHT:
Magnolia campbellii;
Magnolia delavayi.

Characters of similar families: Theaceae: stipules absent, leaf margin usually toothed with glands; stamens basifixed; fruits rarely indehiscent or fleshy. **Annonaceae:** primarily tropical family, stipules absent, fruit aggregate of berries that fuse to a fleshy receptacular axis, indehiscent.

Trees or **shrubs**, evergreen or deciduous. **Buds** enclosed by hooded stipules often floccose (except *Liriodendron*). **Stipules** large, elongated, soon caducous when buds break, annular scars evident around nodes. **Leaves** simple, alternate, petiolate, entire or lobed (*Liriodendron*) and spirally arranged; leaf blade pinnately veined. **Inflorescence** terminal, solitary, usually large, actinomorphic, hermaphrodite, rarely unisexual, spathaceous bracts 1 or more basal to tepals; tepals numerous, free, whorled or spiral, white, cream or pink; stamens usually at basal part of receptacle, numerous, free, spirally arranged around enlarged receptacle; anthers dehisce longitudinally leaving scars; filaments short; gynoecium apocarpous, superior, 2 to numerous, spirally arranged ovaries/carpels placed centrally above enlarged receptacle, each unilocular with one terminal style. **Fruit**, mature carpels usually dehiscing along dorsal and/or ventral sutures or indehiscent (*Liriodendron*), samaroid, and adnate to seed endotesta. **Seeds** usually large, red and fleshy, pendulous on a filiform elastic funiculus exserted from mature carpels (except *Liriodendron*); endospermic, endosperm usually oily, embryo well differentiated, usually very small, larger in *Liriodendron*.

Literature: Bittrich *et al.* (1993); Magnolia Society (2021).

Magnoliaceae is an early diverged lineage within the Magnoliids, and consists of two genera, *Magnolia* and *Liriodendron*. More than 300 species that are disjunctly distributed in the temperate to subtropical zones of Asia, eastern N America and S America.

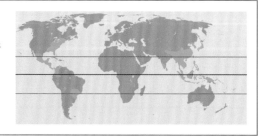

HABIT

Deciduous tree: **1** *Magnolia acuminata* and **2** *Liriodendron tulipifera*.

LEAVES

Large deciduous leaves: **3** *Magnolia macrophylla* subsp. *macrophylla*. Unfolding leaf blade with lateral lobes near base: **4** *Liriodendron chinense*.

FRUIT

Immature fruit with stamen scars at base of receptacle: **8** *Magnolia macrophylla*. Numerous follicles with individual seeds: **9** *Magnolia tripetala*.

FLOWERS

Bud enclosed by hooded stipules: **5** *Magnolia amoena*. Terminal, actinomorphic flower: **6** *Liriodendron chinense*. Elongated receptacle, stamens grouped at base, carpels spirally arranged towards the top: **7** *Magnolia macrophylla*.

Lauraceae

Timothy Utteridge

Stipules absent
Leaves simple
Leaves alternate
Leaves entire
Ovary superior

Trees or **shrubs**. **Leaves** alternate or spiral, glaucous below, stipules absent. **Flowers** 3-merous, small, tepals undifferentiated, anthers opening by flaps; ovary superior; pedicel/receptacle often enlarging and enclosing the single-seeded fruit.

LEFT TO RIGHT:
Litsea cubeba;
Lindera umbellata.

Characters of similar families: **Annonaceae**: flowers large, stamens numerous, anthers linear, fruit compound. **Convolvulaceae** (*Cuscuta*): flowers are needed: fused corolla with 5 lobes (*Cassytha* with free tepals in 3s). **Fagaceae**: stipules present, flowers tiny, wind pollinated, fruit a single-seeded nut subtended by a woody cupule. **Magnoliaceae**: stipules present, petals large and showy, fruit many-seeded.

Trees or **shrubs** (*Cassytha* a twining parasite); cut parts often aromatic. **Sap** absent. **Hairs** where present, simple. **Stipules** absent. **Leaves** simple; usually alternate, rarely subopposite (especially in *Cinnamomum*) or apparently whorled; margins entire, leaves sometimes lobed (e.g. *Sassafras*); often coriaceous, young leaves sometimes red, glaucous below; tripli- or trinerved, lower venation orthogonally arranged; aromatic when crushed. **Inflorescences** axillary, occasionally pseudoterminal; usually branched: thyrsoid or pseudo-umbellate with several umbels of flowers; often enveloped in bracts (*Litsea*, *Lindera* and *Neolitsea*). **Flowers** bisexual and/or unisexual, actinomorphic, very small, yellowish, greenish or white; 3-merous (rarely 2-merous), 2 equal whorls of 3 tepals each; 3 whorls of fertile stamens, 1 inner whorl of sterile staminodes; anthers 2- or 4-locular opening by flaps (valves); ovary unicarpellate, superior, with a single ovule. **Fruit** a berry or drupe with the receptacle or pedicel enlarged surrounding the base of, or entirely enclosing, the fruit. **Seed** 1.

Literature: Rohwer (1993); van der Werff (1991, 2001); van der Werff & Richter (1996); Verdcourt (1996).

A pantropical group with approximately 55 genera and at least 2,500 species. Generic delimitation is problematic, and many taxa are in need of revision. The family is an important component of tropical forests, but its diversity extends into temperate regions; for example, there are ca. 140 species in Australia and ca. 450 species in China. The Mediterranean species *Laurus nobilis* has been introduced into many regions of the world as a spice, but is also used in soap manufacture.

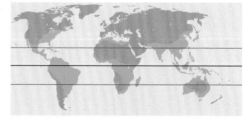

HABIT

Trees and shrubs: **1** *Umbellularia californica* and **2** *Neolitsea sericea*. **3** *Laurus nobilis* planted as an ornamental and trimmed to a cone shape.

LEAVES

Leaves simple: **4** *Laurus nobilis*. Alternate, appearing whorled: **5** *Neolitsea sericea*. Lobed: **6** *Lindera triloba*. Young leaves reddish: **7** *N. sericea*. Undersides glaucous: **8** *N. sericea*.

FLOWERS AND FRUIT

Inflorescences axillary: **9** *Laurus nobilis*. Flowers small, yellow: **10** *Lindera benzoin*. Fruit a drupe: **11** *Cinnamomum camphora* and **12** *Lindera glauca*.

Araceae

Anna Haigh

Herbs
Underground organ often rhizome or tuber
Spathe and spadix present
Fruits fleshy

Herbs, often seasonally dormant geophytes or aquatics, sometimes floating aquatics. **Leaves** petiolate, with a petiole sheath, blade with midrib compound usually with pinnate branches. **Inflorescence** a spathe and spadix, no other bracts. Fruit a berry.

LEFT TO RIGHT:

Arum hygrophilum;
Arisarum simorrhinum: note the tuber and the inflorescence at the base of the leaves.

Characters of similar families: Orchidaceae: flowers large and showy, fruit leathery, dehiscent, seeds numerous, tiny, dust-like. Salviniaceae: leaves are papillate on the aerial surface. Smilacaceae: tendrils usually present, flowers arranged in an umbel.

Herbs, usually seasonally dormant, terrestrial, aquatic, or floating aquatic. **Underground parts** very variable, often a tuber or rhizome. **Hairs** usually lacking, sometimes with trichomes, scales, prickles or warty outgrowths. **Leaves** usually divided into a blade, petiole and petiole sheath; often leathery; blades very variable in shape, entire to deeply lobed or compound, often variegated; midrib compound, primary venation usually pinnately branched but sometimes pedate, arcuate or parallel, secondary venation reticulate or parallel-pinnate. **Inflorescences** scapose, with a spathe (a solitary, specialised bract) and spadix (a dense spike of small flowers), no other bracts. In Lemnoideae (duckweeds), which flower very rarely, the inflorescence is within a minute pouch or pouches. **Flowers** sessile, very small, bi- or unisexual, when unisexual female below male on spadix. **Fruit** a berry. **Seed** very variable, not dust-like.

Literature: Bown (2000); Li & Boyce (2010); Mayo *et al.* (1997); Thompson (2000).

144 genera, ca. 4,000 species, worldwide. Large genera in temperate areas include *Arum* and *Biarum* (Mediterranean), *Arisaema* (Asia and North America) and *Zantedeschia* (Southern Africa, *Z. aethiopica* is invasive in several countries). The floating aquatic *Lemna* is the most widespread genus. Many genera are horticulturally important. The tubers of several genera are eaten throughout the world.

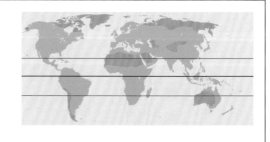

HABIT

Understorey herbs: **1** *Ambrosina bassii* and **2** *Lysichiton americanus*. Floating aquatics: **3** *Lemna minor*. Leaves various: **4** *Arisarum proboscideum* and **5** *Dracunculus vulgaris*. Tuber: **6** *Arum* sp.

INFLORESCENCES

Spathe and spadix diversity: **7** *Arum creticum*, **8** *Dracunculus vulgaris*, **9** *Calla palustris* and **10** *Arisaema filiforme*.

FRUIT

Berries: **11** *Arisaema tortuosum*, **12** *Arum maculatum* and **13** *Arisarum proboscideum*.

Melanthiaceae

Anna Trias-Blasi

Leaves simple
Flowers usually actinomorphic
Flowers bisexual or unisexual
Ovary superior

Flowers often 3-merous, sometimes 4–10(–12)-merous. **Anthers** mostly extrorse. **Ovary** superior with 3 distinct styles. **Fruit** often a capsule.

LEFT TO RIGHT:
Chamaelirium japonicum;
Chamaelirium japonicum: note basal rosette.

Characters of similar families: Colchicaceae: usually with corms, stem sometimes erect and simple, flowers always 6-merous, seeds subglobose, ovoid to subangular. Liliaceae: tepals spotted, stamens always 6, style solitary.

Perennial **herbs**, usually hermaphroditic or andromonoecious, rarely gynodioecious or androdioecious, rarely dioecious. **Underground organ** a rhizome or a bulb, rarely a corm. Aerial stem erect, simple. **Leaves** deciduous or evergreen, spirally arranged, all cauline, or basal ones large and cauline ones small, or all inserted in a basal rosette, or all in a pseudo-whorl at the stem apex, bifacial, sessile or petiolate, linear to broadly elliptical, oblanceolate to obovate or spathulate, rarely setaceous, sometimes sheathing basally. **Inflorescence** terminal, often a raceme or flowers solitary, sometimes a panicle, a spike or umbel-like, glabrous or pubescent. **Flowers** usually hypogynous, rarely half-epigynous or epigynous, often 3-merous, sometimes 4–10(–12)-merous, usually actinomorphic, rarely zygomorphic, bracteate or ebracteate; tepals in two similar whorls or in two separate whorls, usually free, rarely connate basally, filiform to oval or spathulate, white, green, yellow, pink, red, maroon, lilac or dark purple-brown, erect, ascending, spreading or reflexed, persistent, marcescent or caducous, rarely clawed at the base; stamens in 2(–6) whorls, free or inserted at the tepal base, filaments filiform to subulate or flat, anthers linear, lanceolate, suborbicular, obcordate, cordate or hippocrepiform, basifixed or dorsifixed, sometimes with prolonged connective, with distinct or confluent thecae, dehiscing with slits or valves, often extrorse, sometimes latrorse or introrse; ovary superior, composed of partly connate to fully fused carpels, often with 3 distinct styles. **Fruit** often a capsule, sometimes baccate, septicidal, loculicidal, ventricidal or rupturing irregularly due to the combined effect of seed enlargement and breakdown of the fruit wall. **Seeds** linear to ellipsoidal, sometimes flat or angular, sometimes slightly curved, sometimes broadly winged, sometimes appendaged at both ends, sometimes provided with an aril or a scarlet sarcotesta.

Literature: Christenhusz *et al.* (2017); Tamura (1998b); Trias-Blasi *et al.* (2017); Zomlefer *et al.* (2006).

Melanthiaceae contains ca. 17 genera and 180 species, mainly distributed in temperate regions of the northern hemisphere, particularly in N America and E Asia. The circumscription of the family has changed substantially. First, they were placed in the family Liliaceae and later, some genera in the Trilliaceae. The current understanding of Melanthiaceae is based mainly on molecular evidence.

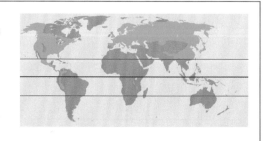

HABIT AND LEAVES

Leaves spirally arranged, basal ones large and cauline ones small, 8-merous flower, trifid style: **1** *Paris incompleta*. Leaves basal and cauline: **2** *Veratrum nigrum*.

FLOWERS

Racemose inflorescence (spike): **3** *Ypsilandra thibetica*. 6-merous flowers: **4** *Trillium grandiflorum* and **5** *Veratrum nigrum*. Flower with 8 tepals and 12 stamens: **6** *Paris rugosa*. 12-merous flower: **7** *Trillium recurvatum*.

Alstromeriaceae

Martin Xanthos

Swollen storage roots present
Leaves alternate
Inflorescence umbellate
Ovary inferior
Fruit a capsule

Rhizomatous **herbs** or **vines**. Leaves simple and alternate or arranged in a basal rosette. **Inflorescence** terminal, few- to many-flowered, often umbel-like. **Flowers** bisexual, usually actinomorphic. **Ovary** inferior with 3 carpels. **Fruit** usually a capsule.

LEFT TO RIGHT: *Luzuriaga marginata*; *Bomarea ovata*: note alternate leaves and twisted leaf bases; *Alstroemeria philippii*: note remnants of capsular fruits.

Characters of similar families: Amaryllidaceae: usually bulbous, inflorescence scapose, pseudoumbellate and enclosed by spathaceous bracts, usually 2. **Asparagaceae**: leaves in a rosette, inflorescence racemose. **Colchicaceae**: corms often tunicated or bulb-like, ovary superior. Iridaceae: corms present, leaves isobifacial, stamens 3.

Perennial **herbs** with rhizomes or **twiners** with swollen storage roots (annual in *Alstroemeria graminea*), some species epiphytic. **Leaves** sessile, simple, alternate, or in a basal rosette, often more or less twisted at the base, so that the blade is resupinate. **Inflorescences** terminal, bracteate with leaf-like bracts (sometimes scale-like or arranged in a false whorl), few- to many-flowered and often umbel-like, rarely solitary. **Flowers** bisexual, actinomorphic or zygomorphic. **Tepals** 2-whorled, petaloid, the inner whorl nectariferous at the base; stamens free, in 2 whorls of 3; ovary inferior with 3 carpels, 3-locular (1-locular in some *Bomarea*), placentation parietal or axile; style filiform, stigma 3-branched. **Fruit** a dry, leathery or rarely fleshy capsule (rarely a berry); dehiscence loculicidal (explosive in *Alstroemeria*). **Seeds** globose, often with an orange-red aril; embryo cylindrical; endosperm oily.

Literature: Christenhusz *et al.* (2017); Heywood *et al.* (2007); WCSP (2020).

Alstroemeriaceae includes 4 genera and 253 species. Mainly confined to Central and South America, also occurring in Australia, New Zealand and the Greater Antilles. Several species of *Alstroemeria* are of great economic importance in horticulture, whereas species of *Bomarea*, particularly *B. edulis*, are cultivated for their starchy edible roots.

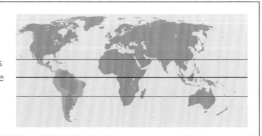

UNDERGROUND PARTS AND LEAVES

Swollen storage roots: **1** *Alstroemeria pulchella*. Leaves arranged in a rosette: **2** *A. pulchella*.

INFLORESCENCES

Actinomorphic and zygomorphic petaloid flowers: **3** *Alstroemeria* sp., **4** *A. aurea*, **5** *A. psittacina*, **6** *A. pulchella*, **7** *A. pelegrina*, **8** *A. paupercula* and **9** *A. ligtu*.

FRUITS AND SEEDS

Dry leathery capsules and globose fruits: **10** *Alstroemeria diluta* (fruits) **11** *A. aurea* (capsules) and **12** *A. pulchella* (capsules).

1 mm

Colchicaceae

Anna Trias-Blasi

Corms present
Tepals 6
Stamens 6
Styles divided in 3 styluli
Ovary superior

Usually corms. Leaves usually sessile. Tepals 6.
Stamens 6. **Styles** often divided into 3 free styluli.
Ovary superior. **Fruit** usually a dry or fleshy capsule.

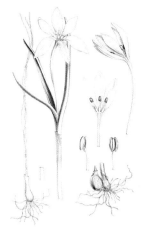

LEFT TO RIGHT:
Colchicum montanum; Colchicum graecum.

Characters of similar families: Amaryllidaceae: 2 spathaceous bracts, inflorescence a pseudoumbel, ovary inferior, undivided styles. **Asparagaceae:** leaves normally in basal rosettes, sometimes one spathaceous bract, pedicels articulated. **Iridaceae:** leaves isobifacial, stamens 3, ovary inferior. **Liliaceae:** usually bulbs, leaves mostly cauline, styles never divided into 3 free styluli. **Melanthiaceae:** rarely with corms, stem always erect and simple, flowers 4–10(–12)-merous, seed linear to ellipsoid. erect and simple, flowers 4–10(–12)-merous, seed linear to ellipsoid.

Perennial herbs with rhizomes or corms, leaves and flowers annual; roots sometimes tuberous. **Stem** erect, simple or branching, sometimes scandent or reduced to a short underground portion. **Leaves** cauline, distichous, alternate or subopposite to verticillate, sessile or subpetiolate, usually sheathing, bifacial; venation parallel with often a distinct midrib, rarely with reticulate secondary venation. **Inflorescence** a terminal raceme or cyme, sometimes umbellate, spike-like or flowers solitary. **Flowers** bisexual, rarely unisexual, actinomorphic to slightly zygomorphic; tepals 6(–12), equal or somewhat unequal, free or fused in lower half, sometimes spotted or variegated, usually with perigonal or androecial nectaries, caducous or persistent; stamens 6, dorsifixed, dehiscing extrorsely by longitudinal slits, rarely latrorsely to introrsely; ovary superior 3(–4)-locular, completely or partially fused with axile placentation; ovules few to many per locule; styles 1–3(–4), free or partly united styluli, styles and stigmas often minute. **Fruit** a dry or fleshy septicidal or loculicidal capsule, rarely berry-like. **Seeds** subglobose or ovoid to subangular, with or without a dry strophiole, sometimes with a fleshy aril.

Literature: Nordenstam (1998); Christenhusz *et al.* (2017).

Colchicaceae has ca. 280 species in 15 genera distributed in temperate to tropical zones of Africa, Europe, Asia, Australia and N America. The family is characterised by the presence of colchicine, a medication used to treat gout amongst other diseases.

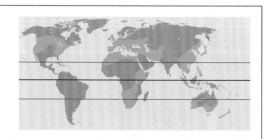

HABIT AND LEAVES

Stem reduced to a short underground portion, leaves sessile: **1** *Colchicum vanjaarsveldii* and **2** *C. atticum*. Stem erect, leaves sessile: **3** *Disporum uniflorum*.

FLOWERS

6 reflexed tepals, 6 stamens, style, superior ovary: **4** *Gloriosa superba*. 6 tepals, 6 stamens, 3 styluli: **5** *Colchicum autumnale* and **6** *C. palaestinum*.

FRUIT

Capsule: **7** *Iphigenia indica*. Capsule with seeds: **8** *Gloriosa superba*. Berry: **9** *Disporum cantoniense*.

45

Smilacaceae

Anna Haigh & Paul Wilkin

Climbing plants
Underground storage organ – rhizome
Inflorescences umbellate
Flowers unisexual
Ovary superior

Climbers, with paired tendrils inserted at each node or petiole base. **Leaf** short-petiolate, primary venation parallel with veins fusing at the tip. **Inflorescence** umbellate, unspecialised 'monocot' flowers, usually unisexual, ovary superior. **Fruit** a berry.

LEFT TO RIGHT:
Smilax gaudichaudiana;
Smilax maritima: note tendrils, umbellate infructescences and spiny stems.

Characters of similar families: Araceae: inflorescence consists of a spathe and spadix. **Dioscoreaceae**: herbaceous climbers lacking tendrils, ovary inferior, fruits winged. **Asparagaceae**: lacking tendrils when climbing, highly variable but inflorescences often racemose, fruit often capsular. **Vitaceae**: tendrils and inflorescences leaf opposed, leaves often toothed. **Menispermaceae**: tendrils lacking, fruits with a strongly curved endocarp

Climbers (less often herbs or shrubs), often armed with prickles or unarmed, climbing with tendrils (modified from the petiole base). **Underground parts** rhizomatous. **Hairs** present or absent. **Stipules** often present at each node or petiole base as a pair of narrow to inflated lobes bearing the tendrils. **Leaves** alternate, opposite or verticillate, entire, usually coriaceous, prominently 3–7-veined with a midrib present, all main veins originating at blade base, parallel and joining at the tip, higher-order veins finely reticulate; petiole present. **Inflorescences** usually umbellate (rarely racemose or spicate), simple or compound, usually axillary. **Flowers** unisexual (plants dioecious), rarely bisexual, usually small, greenish to yellow, with 6 free or fused tepals, usually 6 stamens, rarely 3–18, anthers 1-locular; pistillate flowers with staminodes, ovary superior, 3-locular, 1–2 ovules per locule. **Fruit** a red, black or purple berry. **Seeds** 1–3.

Literature: Chen & Koyama (2000); DeFilipps (1980); Holmes (2002).

Only one genus (*Heterosmilax* is now considered part of *Smilax*), with ca. 260 species. *Smilax* is mostly pantropical with fewer species in temperate areas; there are only two species in Africa (one in southern Africa). The group has a number of medicinal uses, *Smilax aristolochiifolia* was traditionally an antisyphilitic, but is today a tonic and a flavouring for drinks.

HABIT

Climbers with tendrils and stipule-like structures: **1** *Smilax china*. Leaves simple with veins originating at the base and meeting at the apex: **2** *S. aspera*. Scrambling habit: **3** *Smilax megalantha*.

FLOWERS AND FRUITS

Female flowers: **4** *Smilax china* and **5** *S. aspera*. Male flowers: **6** *S. auriculata*. Mature fruits: **7** *S. aspera*.

Liliaceae

Anna Trias-Blasi

Bulbs present
Flowers large
Nectaries present
Stamens 6
Ovary superior

Herbs. Bulbs or **rhizomes. Leaves** mostly cauline. **Flowers** often large. **Tepals** 6, often spotted with nectaries. **Stamens** 6. **Ovary** superior.

RIGHT:

Lilium martagon: note cauline leaves, reflexed tepals and 6 stamens.

Characters of similar families: Amaryllidaceae: inflorescence a pseudoumbel; 2 spathaceous bracts. **Colchicaceae:** usually corms, leaves basal or cauline, styles often divided into 3 free styluli. **Iridaceae:** leaves cauline, sheathing at the base, isobifacial, stamens 3, branching styles; ovary inferior. **Melanthiaceae:** tepals unspotted, stamens not always 6, free styles or fused with free stigma. **Asparagaceae:** leaves normally in basal rosettes, sometimes one spathaceous bract, pedicels articulated.

Perennial **herbs**. Storage organs usually bulbs, sometimes rhizomes. **Leaves** mostly cauline, sometimes basal, spiral or (in *Lilium* and *Fritillaria* species) whorled, rarely petiolate, simple, and parallel-veined. **Inflorescence** a terminal raceme, a solitary flower, or rarely an umbel. **Flowers** often large, bisexual, actinomorphic or zygomorphic, pedicellate, bracteate or not, hypogynous. **Tepals** biseriate, 3+3 sometimes spotted or striate. **Stamens** are 3+3, whorled, distinct and free. **Anthers** are peltately attached to the filament or pseudobasifixed, and longitudinally dehiscent. **Ovary** syncarpous, superior, 3 carpels and 3 locules. **Style** solitary; stigmas 3, tri-lobed or with 3 crests; axile placentation. **Nectaries** perigonal, present at the tepal bases. **Fruit** loculicidal, septicidal, or irregularly dehiscent capsules or a berry. **Seeds** flat discoid or ellipsoid, the endosperm with aleurone and fatty oils.

Literature: Chen *et al.* (2000); Simpson (2010); Tamura (1998a).

Liliaceae s.s. has ca. 700 species in 15 genera, with a mostly north temperate distribution. In the past, the family was treated as a large assemblage (Liliaceae s.l.), which has since been broken up into numerous segregate families. It has economic value as several taxa, including *Lilium* and *Tulipa*, are ornamental cultivars.

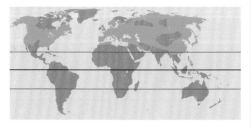

HABIT AND LEAVES

Habit, basal leaves: **1** *Tulipa saxatilis*. Habit, upper and lower leaves: **2** *Fritillaria imperialis*. Habit, cauline leaves, 6 spotted tepals, 6 stamens: **3** *Lilium pensylvanicum*.

FLOWERS AND FRUITS

Tepals reflexed and spotted, 6 stamens: **4** *Lilium martagon*. Tepals erect: **5** *Tulipa linifolia*. 6 stamens and superior ovary: **6** *Gagea lutea*. Nectaries: **7** *Fritillaria* sp. Capsules: **8** *Fritillaria ruthenica* and **9** *Lilium candidum*.

49

Orchidaceae

André Schuiteman

Flowers with a labellum
Sexual organs fused
Pollinia
Ovary inferior
Microseeds

Herbs, in temperate regions mainly geophytes.
Flowers zygomorphic; one tepal, the labellum, is usually quite different from the other five; sexual organs fused into a column; pollen clumped into pollinia; dust-like seed.

LEFT TO RIGHT:
Anacamptis morio, showing habit, floral details, pollinia and seeds;
Caladenia filamentosa (Tasmania);
Chloraea magellanica (Patagonia).

Characters of similar families: Zingiberaceae: plants aromatic, leaves with open sheaths and ligule, petaloid staminodes, no pollinia, seed with endosperm. **Asparagaceae:** flowers actinomorphic, 3 or more stamens, no pollinia, seed with endosperm. **Araceae:** inflorescence a spadix with a single, basal bract, no other bracts.

Herbs, mycotrophic at least as seedlings (some species mycoheterotrophic as mature plants), initially forming a protocorm (a leafless and rootless body); terrestrial, lithophytic, or epiphytic; temperate species often geophytes. **Roots** usually with a velamen (outer zone of dead cells), often thick, little-branched and sparse. **Hairs** present or absent. **Leaves** entire, spirally arranged or distichous, often with basal sheath, sometimes reduced to scales. **Inflorescences** racemose or sometimes paniculate, often with a single flower only. **Flowers** usually resupinate (the labellum situated below the column), zygomorphic, minute to large; 6 tepals in 2 whorls, petaloid; outer tepals (sepals) 3; inner tepals 3, comprising 2 petals and a usually very different labellum (lip); tepals sometimes connate; lip or sepals sometimes spurred; sexual organs fused into a column; anthers 1, 2 or very rarely 3; pollen usually clumped into pollinia (2, 4, 6 or 8 small bodies), rarely gel-like or powdery; ovary inferior with numerous (often >1,500) tenuinucellate ovules per carpel. **Fruit** usually a capsule splitting open along 1–6 zones of dehiscence, rarely an elongate berry. **Seeds** minute (microseeds), consisting of an embryo surrounded by a testa, lacking endosperm and phytomelan.

Literature: Jones (1988); Kühn *et al.* (2019); Linder & Kurzweil (1999); Pridgeon *et al.* (1999–2014).

Approximately 750 genera, 25,000–30,000 species. Cosmopolitan but predominantly tropical and subtropical, with >75% of the species epiphytic. Usually indicative of oligotrophic and biodiverse habitats. Often with deceptive pollination based on floral mimicry. Tubers of temperate species sometimes eaten (e.g., *salep* in Turkey), often harvested non-sustainably.

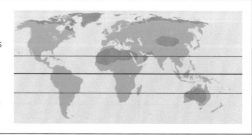

HABIT
Single-flowered inflorescence: **1** *Calypso bulbosa*. Many-flowered inflorescence: **2** *Malaxis monophyllos*. Mycoheterotrophic orchid with non-resupinate flowers and without chlorophyll: **3** *Epipogium aphyllum*.

FLOWERS
Insect-mimicking orchid: **4** *Ophrys speculum*. Slipper orchid with 2 anthers and connate lateral sepals: **5** *Cypripedium reginae*. Lip and column partly fused: **6** *Calanthe lamellosa*. Dorsal sepal spurred: **7** *Disa aurata*.

REPRODUCTION
Column: **8** *Epipactis palustris* and **9** *Platanthera bifolia*. Andrena bee escaping while pollinating the flower that had trapped it: **10** *Cypripedium calceolus*. Fruiting orchid dispersing its dust-like seed: **11** *Epipogium roseum*.

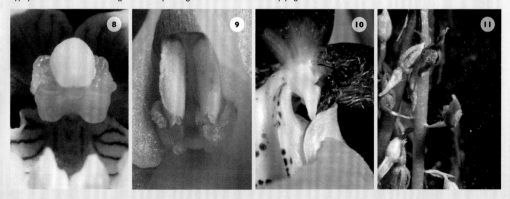

Iridaceae

Anna Trias-Blasi

Leaves isobifacial
Stamens 3
Style branched
Ovary inferior
Fruit a capsule

Usually **herbs** with underground storage organs.
Leaves usually sheathing at the base, usually isobifacial.
Stamens 3. **Ovary** inferior.

LEFT TO RIGHT:
Crocus salzmannii: bulb, leaves sheathing at the base, 3 styles;
Iris colchica: note isobifacial leaves, 6 tepals and inferior ovary.

Characters of similar families: Amaryllidaceae: leaves cauline, usually not isobifacial, stamens 6, usually bulbs. **Colchicaceae**: leaves cauline, usually not isobifacial, stamens 6, ovary superior, usually corms. **Liliaceae**: leaves cauline, usually not isobifacial, stamens 6; ovary superior, bulbs. **Asparagaceae**: leaves not isobifacial, stamens usually 6, ovary usually superior, style not branched. **Asphodelaceae**: leaves rosulate, not isobifacial, stamens 6, ovary superior, style not branched.

Usually **perennial herbs** or with an annual shoot system. Underground storage organ a bulb, a corm or a rhizome. Some taxa have a hairy **indumentum**. **Leaves** simple, basal and cauline, sheathing at the base, blades isobifacial with their edge orientated towards the stem or bifacial with the lower surface facing the stem. Flowering stem aerial or subterranean at anthesis, simple or branched. **Inflorescences** spikes or monochasial cymes, sometimes reduced to a single flower; scape erect, cylindrical or flattened. **Flowers** bisexual, actinomorphic or zygomorphic, usually large and showy, with 2 petaloid whorls of 3 tepals each; stamens usually 3, anthers extrorse; ovary inferior, trilocular, multiovulate with axile placentation; styles 3 sometimes with bifid branches. **Fruit** a capsule, carrying 1–many seeds, usually dehiscent. **Seed** often large, globose to angular or discoid, brown to black; endosperm hard with reserves of hemicellulose, oil and protein.

Literature: Avila (2012); Goldblatt (1993, 1996, 1998); Goldblatt & Manning (2008).

The family comprises ca. 70 genera and 2,000 species. It is mostly pantropical and most diverse in southern Africa, but some taxa are distributed partially or completely in temperate areas. Many genera, such as *Iris* and *Crocus*, contain species that are important ornamentals. The stigmas of *Crocus sativus* provide saffron.

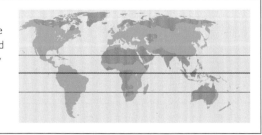

HABIT AND LEAVES

Habit: **1** *Iris foetidissima*. Isobifacial leaves: **2** *Iris* sp. Leaves sheathing at the base: **3** *I. aucheri*.

INFLORESCENCE AND FLOWERS

Inflorescence: **4** *Dierama pulcherrimum*. Flowers with differentiated tepal whorls and branched style: **5** *Iris wattii*. Three stamens and branched styles: **6** *Crocus* sp. Undifferentiated tepal whorls, 3 stamens and branched style: **7** *Crocus boryi*.

FRUITS

Capsules: **8** *Libertia ixioides*. Capsule with seeds: **9** *Iris foetidissima*.

Asphodelaceae

Anna Haigh

Underground organs various
Stamens 6
Inflorescence various
Ovary superior
Seeds often black

Herbs or with woody stems. **Leaves** rosulate or distichous, basal or cauline. **Pedicels** often articulated. **Fruit** a capsule or berry. **Seeds** often black.

LEFT TO RIGHT:
Eremurus species: note fleshy roots;
Bulbinella rossii.

Characters of similar families: Amaryllidaceae: usually bulbous, inflorescences scapose, pseudoumbellate, enclosed by fleshy spathaceous bracts in bud. **Asparagaceae:** inflorescences usually racemose, also spicate, paniculate or cymose; Asphodelaceae and Asparagaceae have few morphological characters to distinguish them, comparison at the subfamily or generic level may be more helpful. **Iridaceae:** usually with isobifacial leaves, stamens 3, ovary inferior, styles branched. **Liliaceae:** bulbous herbs with cauline leaves, flowers solitary or more rarely arranged in a thyrse, raceme or umbel.

Herbs or **arborescent** (rarely **climbing**), perennial or annual. Underground organs usually rhizomes or with tuberous or fibrous roots. **Hairs** uncommon. **Leaves** rosulate or distichous, basal or cauline. Usually ovate to linear with a sheathing base, rarely petiolulate or succulent with spiny margins; venation parallel. **Inflorescences** bracteate, usually an axillary or terminal panicle, raceme or cyme or a combination of these, occasionally flowers solitary. **Flowers** bisexual, pedicels often articulated, actinomorphic to somewhat zygomorphic. Tepals free and spreading or occasionally fused and tubular. **Stamens** 6 basifixed or dorsifixed, introrse or extrorse, filaments glabrous or papillose. **Ovary** superior, 3-locular; style usually filiform; stigma usually minute, occasionally capitate or trilobed. **Fruit** usually a loculicidal capsule. Rarely a berry, schizocarp or nut-like and indehiscent or rupturing. **Seeds** 3 to numerous, usually black.

Literature: Bedford *et al.* (2020); Chase *et al.* (2009); Moore & Edgar (1970); Navas Bustamante (1973); Tutin *et al.* (1980).

Asphodelaceae as defined by APG IV is morphologically heterogenous, and it includes many taxa traditionally placed in separate families, making the family difficult to recognise morphologically. A cosmopolitan family of 40 genera and ca. 1,000 species made up of three subfamilies: Asphodeloideae (predominantly southern Africa), Hemerocallidoideae (predominantly Asia to Australia) and Xanthorrhoeoideae (Australia).

HABIT

Basal leaves: **1** *Hemerocallis lilioasphodelus* and **2** *Asphodelus tenuifolius*. Clump-forming: **3** *Phormium tenax*. Arborescent: **4** *Xanthorrhoea quadrangulata*.

INFLORESCENCES AND FLOWERS

Inflorescence racemose: **5** *Asphodelus tenuifolius* and **6** *Kniphofia northiae*. Cymose: **7** *Hemerocallis fulva*. Paniculate: **8** *Phormium colensoi*.

FRUIT

Berries: **9** *Dianella ensifolia*. Capsules (immature): **10** *Kniphofia northiae*, Capsules (mature): **11** *Xanthorrhoea bracteata*.

Amaryllidaceae

Martin Xanthos

Underground parts bulbous
Leaves in a basal rosette
Inflorescence scapose
Inflorescence with 2 spathaceous bracts
Ovary usually inferior

Bulbous herbs, rarely rhizomatous. **Leaves** in a basal rosette. **Inflorescence** with 2 spathaceous bracts, pseudo-umbellate or flowers solitary. **Perianth** of tepals usually in 2 series, corona sometimes present.

LEFT TO RIGHT:

Nerine sarniensis;

Narcissus cyclamineus x *pseudonarcissus*: note leaf rosette and scapose inflorescence;

Leucojum ionicum: note bulbous base and inferior ovary.

Characters of similar families: **Asparagaceae**: woody rhizome or corm present, inflorescence racemose, lacking spathaceous bracts or 3 or more bracts, bracts not enclosing floral buds and subtending the inflorescence, flowers on articulated pedicels. **Alstroemeriaceae**: not bulbous, leaf blades resupinate and usually cauline, inflorescence not scapose with spathaceous bracts. **Asphodelaceae**: underground organs various, inflorescence various, ovary superior. **Colchicaceae**: corms, leaves cauline, inflorescence not scapose with spathaceous bracts. **Iridaceae**: corms, leaves isobifacial, stamens 3. **Liliaceae**: naked or tunicate bulbs, leaves cauline, ovary superior.

Herbs, geophytic forming a tunicate bulbous rootstock, rarely rhizomatous. **Leaves** spiral or distichous in a basal rosette, usually linear. **Inflorescences** scapose, pseudoumbellate, subtended by 2 or more spathaceous bracts. **Flowers** 1–many, usually large and showy, bisexual, actinomorphic, consisting of tepals, petaloid, arranged in 2 series, the outer connate below into a tube, rarely free to the base, the inner generally shorter than the outer, sometimes forming a conspicuous corona; stamens 6 opening by longitudinal slits; stigmas capitate, 3-lobed; ovary inferior or superior, usually 3-locular. **Fruit** a capsule, often dehiscent, rarely a berry. **Seeds** globose or subglobose, fleshy and hard, or flattened and winged, usually with a black or brown phytomelanous testa.

Literature: Angiosperm Phylogeny Group (APG) (2009); Chase *et al.* (2009); Wilkin (2015a).

The family is cosmopolitan and comprises 3 subfamilies: Amaryllidoideae (59 genera, c. 850 species) is predominantly tropical but genera such as *Cyrtanthus, Galanthus, Gethyllis, Haemanthus* and *Narcissus* also occur in temperate and warm temperate areas; Allioideae (15 genera, c. 800 species) found mainly in warm temperate and Mediterranean regions; and Agapanthoideae 1 genus (*Agapanthus*) of 9 species from South Africa.

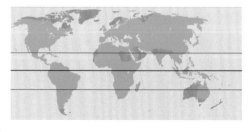

HABIT AND LEAF ARRANGEMENT

Leaves in a rosette: **1** *Allium karataviense*. Distichously arranged: **2** *Crinum campanulatum*. Whorled: **3** *Galanthus ikariae*.

INFLORESCENCES

Scapose, pseudoumbellate:
4 *Agapanthus* 'Purple Cloud'
and **5** *Crinum* 'Grace Hannibal'.

FLOWERS

Showy flowers, 6 petaloid tepals, 6 stamens: **6** *Agapanthus* 'Peter Pan', **7** *Narcissus* sp., **8** *Ammocharis coranica*, **9** *Ipheion uniflorum* and **10** *Galanthus nivalis*.

FRUITS

Capsules, often dehiscent: **11** *Pancratium canariense* and **12** *Allium ursinum*.

Asparagaceae

Martin Xanthos

Leaves usually in rosettes
Inflorescence racemose
Spathaceous bracts if present not 2, not enclosing floral buds
Stamens usually 6
Seeds often black

Herbs or **shrubs**, occasionally arborescent. **Leaves** arranged in a rosette or clustered on short stems. **Inflorescences** often racemose, pedicels often articulated. **Fruit** a capsule or berry. **Seeds** often black.

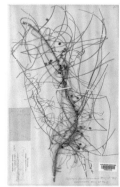

LEFT TO RIGHT:

Muscari anatolicum;

Cordyline murchisoniae: note leaves clustered on the stem and the racemose inflorescence;

Asparagus filicinus: note fruits are a berry.

Characters of similar families: **Amaryllidaceae:** usually bulbous, inflorescence pseudoumbellate and enclosed by spathaceous bracts, usually 2. **Alstromeriaceae:** leaf blades resupinate and usually cauline, inflorescence not scapose with spathaceous bracts. **Asphodelaceae:** inflorescence usually thyrsoid; there are few morphological characters to distinguish this family from Asparagaceae, so comparison at the subfamily or generic level may be more helpful. **Colchicaceae:** corms, leaves cauline, inflorescence not scapose with spathaceous bracts. **Iridaceae:** corms, leaves isobifacial, stamens 3. **Liliaceae:** naked or tunicate bulbs, leaves cauline. **Palmae:** leaves compound, plicate and splitting on folds, leaf base tubular. **Smilacaceae:** climbers with paired tendrils, inflorescence umbellate, flowers unisexual.

Herbs, **shrubs**, or **arborescent** with woody growth via anomalous secondary thickening. **Underground organs** rhizomes, tubers, bulbs or a woody caudex. **Stems** usually herbaceous, sometimes woody (*Cordyline* and *Yucca*). **Leaves** normally arranged in a basal rosette, sometimes cauline (for example, in *Maianthemum* and *Ruscus*), or replaced by modified green stems (phylloclades) (*Asparagus*), or fleshy and spine-tipped in *Agave*. **Inflorescences** usually racemose, also spicate, paniculate or cymose, with 3 or more bracts subtending but not enclosing the inflorescence, or no bracts. **Flowers** often actinomorphic on articulated pedicels, sometimes fragrant, tepals 6, free or fused, stamens usually 6, and inserted at the base of the tepal lobes or on the tube, filaments free, ovary superior or inferior, 3-locular, 1–many seeds per locule, style short with lobed, divided or capitate stigma. **Fruit** a capsule or berry. **Seeds** often black, rounded, subglobular or flattened, sometimes winged.

Literature: Angiosperm Phylogeny Group (APG) (2009); Chase *et al.* (2009); Wilkin (2015b).

A morphologically heterogeneous family of c.150 genera and 2,595 species. The family is cosmopolitan and includes genera that previously belonged to other traditional families. Some of the more notable temperate genera, including introduced taxa, are *Anthericum*, *Asparagus*, *Cordyline*, *Hosta*, *Hyacinthoides*, *Ruscus*, *Scilla* and *Yucca*.

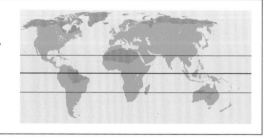

HABIT AND LEAF ARRANGEMENT

Leaves arranged in rosettes: **1** *Cordyline australis*, **2** *Hosta sieboldiana* and **3** *Asparagus mairei*.

INFLORESCENCES

Racemose inflorescences: **4** *Hyacinthoides non-scripta*, **5** *Cordyline australis*, **6** *Dracaena draco* and **7** *Beschorneria yuccoides*.

FRUITS AND SEEDS

Capsules or berries often with black seeds: **8** *Asparagus laricinus*, **9** *Dipcadi viride*, **10** *Maianthemum racemosum*, **11** *Drimia aphylla* and **12** *Bellevalia hermonis*.

Palmae

William J. Baker

Woody plant
Leaf (bi)pinnate or palmate
Leaf plicate, splitting along folds
Leaf base tubular
Ovary superior, 1 ovule per carpel

Woody monocot **trees**, **shrubs** or **climbers**. **Leaves** usually compound, palmate, pinnate or rarely bipinnate, blade plicate, splitting along folds. **Leaf** bases tubular, usually sheathing. **Inflorescences** lateral. **Fruit** a berry or drupe.

LEFT TO RIGHT:
Jubaea chilensis: habit;
Butia eriospatha;
Trachycarpus fortunei.

Characters of similar families: Pandanaceae: linear leaves, flowers complex and not usually trimerous, fruit in heads. **Cyclanthaceae:** leaves not developing from an initially entire lamina. **Asparagaceae** (for example, arborescent *Yucca*, *Cordyline*, *Dasylirion* and *Dracaena*): linear leaves, not compound or plicate. **Musaceae:** non-woody pseudostems arising from corms, leaf entire. **Cycadaceae:** cone-bearing, compound leaves not developing from an initially entire lamina.

Massive to minute woody monocotyledonous plants; **trees**, **shrubs** or **climbers**, sometimes spiny. **Stems** clustered or solitary, erect, creeping or climbing, often massive, sometimes very short and/or subterranean, usually marked with leaf scars, sometimes dying after flowering, aerial stems usually unbranched. **Leaves** compound, less frequently entire; compound leaves derived by splitting of an initially entire lamina, pinnate, bipinnate or palmate. Lamina almost always conspicuously folded. Leaf bases always tubular and sheathing (at least in the bud), often forming a conspicuous tubular crownshaft. **Inflorescences** always lateral, often massive, spicate to paniculate, sometimes aggregated in a mass of inflorescences held above the leaves, resulting in the death of the stem. **Flowers** bisexual or unisexual (monoecious or dioecious); calyx or corolla usually comprising 3 free or fused sepals or petals, often inconspicuous, but some colourful and enlarged; ovary always superior, always no more than one ovule per carpel. **Fruit** a berry or drupe; minute to massive, usually 1–3-seeded, often brightly coloured, sometimes scaly or spiny.

Literature: Baker & Dransfield (2016); Dransfield *et al.* (2008); Henderson (2009); Henderson *et al.* (1995); Meerow (2005).

181 genera, c. 2,600 species. Primarily tropical and subtropical, especially in humid forests, but reaching temperate climates at their northern (southern France) and southern (Chatham Islands) limits. Also reaching high elevations in the tropics (up to 3,600 m in the Andes). An increasing variety of species are cultivated in temperate zones. Widely used by humans for many different purposes.

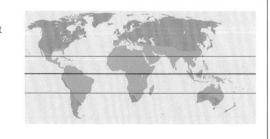

HABIT AND LEAVES

Palmate-leaved palms: **1** *Chamaerops humilis* and **2** *Trachycarpus martianus*. Pinnate-leaved palms: **3** *Phoenix canariensis* and **4** *Rhopalostylis sapida*.

INFLORESCENCES

5 *Chamaerops humilis* and **6** *Phoenix canariensis*.

FLOWERS AND FRUIT

Flowers: **7** *Chamaerops humilis*. Fruit: **8** *C. humilis* and **9** *Butia yatay*.

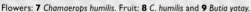

Juncaceae

David A. Simpson & Martin Xanthos

Culms rounded, rarely flattened, without nodes
Flowers small to minute, regular
Perianth of 6 tepals
Fruit a capsule
Seeds 3–many

Rhizomatous or annual **herbs**. Culms mostly terete. **Leaves** mostly linear, glabrous, sometimes hairy. **Flowers** small to minute, regular. **Perianth** of 6 tepals, 2 whorls of 3. **Fruit** a capsule. **Seeds** 3–many.

RIGHT:
Marsippospermum gracile.

Characters of similar families: Cyperaceae: culms 3-sided, leaves usually 3-ranked, flowers minute, perianth reduced to bristles, rarely tepals, or 0, fruit a 1-seeded nutlet. **Poaceae:** culms rounded or flattened and always hollow and noded, leaves 2-ranked, lemma and palea present in floret, fruit a caryopsis. **Restionaceae:** plants mostly dioecious, leaves usually reduced to sheaths, flowers mostly unisexual, minute and regular, perianth of 6 tepals, fruit a 1–3-locular capsule or nutlet.

Herbs, usually rhizomatous, rarely annual. **Culms** erect or rarely procumbent, terete or rarely flattened, without nodes. Leaves linear to filiform, spirally arranged or rarely distichous, glabrous or sometimes hairy, blades sometimes reduced to a bladeless sheath. **Inflorescence** terminal, sometimes pseudolateral, compound, racemose or cymose, in heads or spike-like clusters, rarely a single terminal or lateral flower. **Flowers** small to minute, regular. **Perianth** comprising glume-like tepals. **Tepals** 6, in 2 whorls of 3, 8 mm or more long, acute at apex, ± equal. **Stamens** 6, in 2 whorls of 3, inner whorl sometimes reduced. **Stigmas** 3. **Fruit** a 3–many-seeded capsule, oblong to orbicular or ellipsoid, 3-lobed. **Seeds** sometimes with tail-like appendages, often with distinct sculpturing on surface, whitish to brown or yellowish.

Literature: Balslev (1998); Kirschner *et al.* (2002a–c).

Eight genera and 470 species. Widespread in both northern and southern hemisphere temperate regions, as well as at higher altitudes in the tropics. The largest genera are *Juncus* (331 species) and *Luzula* (123 species). Juncaceae have little or no economic use.

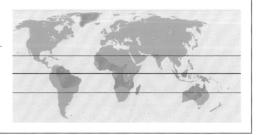

HABIT

Herbs: **1** *Juncus effusus*, **2** *Juncus compressus*, **3** *Juncus rigidus*, **4** *Juncus triglumis* and **5** *Luzula pilosa*.

FLOWERS

6 *Luzula oligantha*, **7** *Juncus coriaceus*, **8** *Juncus capitatus* and **9** *Luzula nivea*.

Cyperaceae

David A. Simpson & Martin Xanthos

Underground rhizomes or stolons
Culms 3-sided, rarely noded
Leaves 3-ranked
Flowers minute
Fruit a minute, 1-seeded nutlet

Rhizomatous, stoloniferous, or annual **herbs**. **Culms** 3-sided. **Leaves** grass-like. **Inflorescence** with 1–many partial inflorescences (spikelets or spicoids). **Flowers** minute. **Fruit** a hard, 1-seeded, nutlet, sometimes enclosed by a sac-like utricle.

LEFT TO RIGHT:
Cyperus rotundus;
Carex halophila.

Characters of similar families: Juncaceae: stems rounded, flowers mostly hermaphrodite, minute and regular, perianth of 6 tepals, fruit a capsule with 3–many seeds. **Poaceae:** culms rounded or flattened and always hollow and noded, leaves 2-ranked, lemma and palea present in floret, fruit a caryopsis. **Restionaceae:** plants mostly dioecious, leaves usually reduced to sheaths, flowers mostly unisexual, minute and regular, perianth of 6 tepals, fruit a 1–3-locular capsule or nutlet.

Herbs, rhizomatous to stoloniferous, or annual. **Culms** mostly 3-sided, solid, rarely noded. **Leaves**, mostly 3-ranked toward base of culm, usually linear, grass-like, sometimes broader with a pseudopetiole, sometimes reduced to a sheath; ligule often present. **Involucral bracts** 1–several, leaf-like or glume-like. **Inflorescence** terminal, sometimes pseudolateral, simple or branched, umbel-like or paniculate, with 1–many partial inflorescence units (spikelets or spicoids). **Spikelets** comprising 1–many, spiral or 2-ranked glumes, each subtending 1 minute flower or sterile; spicoids comprising a female flower, 2–12 scale-like bracts, the outermost 2 bracts opposite and keeled, the spicoid subtended and usually hidden by a glume-like bract. **Perianth** absent or bristle- or scale-like. **Stigmas** 2–3. **Fruit** a hard, 1-seeded, 2–3-sided nutlet, smooth or patterned, sometimes enclosed by a sac-like utricle (perigynium) or with a basal cup-like disk.

Literature: Dai *et al.* (2010); Goetghebeur (1998); Govaerts *et al.* (2007); Semmouri *et al.* (2019); Simpson & Inglis (2001).

91 genera and c. 5,500 species. Worldwide except Antarctica. Particularly common in open wetlands but also found in a range of other habitats, including grasslands and forests. *Carex* is the largest temperate genus. Uses include basketry, construction, food, matting and medicine; some species are grown as ornamentals.

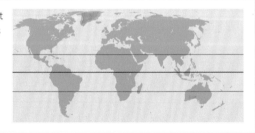

HABIT

Herbs with linear, grass-like leaves: **1** *Carex pendula*, **2** *Bolboschoenus maritimus*, **3** *Cladium mariscus*, **4** *Rhynchospora alba* and **5** *Eriophorum angustifolium*.

INFLORESCENCES

Inflorescences comprising several partial inflorescences of spikes or spikelets: **6** *Carex nigra*, **7** *Cyperus fuscus* and **8** *Cyperus eragrostis*.

CAREX UTRICLES

9 *Carex demissa* and **10** *Carex grayi*.

Restionaceae

Laura Jennings

No underground storage organ
Flowers small and reduced
Ovary superior
Fruits dry

Habit dioecious, rush-like plants. **Culms** usually solid, photosynthetic. **Leaves** reduced to sheaths. **Flowers** tiny, grouped in spikelets, wind-pollinated, 6 tepals in 2 whorls.

LEFT TO RIGHT:
Chordifex monocephalus (left) and
Lepyrodia muelleri (right);
Restio virgeus.

Characters of similar families: Poaceae: leaf blades present at the nodes, culms hollow, tepals 3 or fewer, flowers bisexual, fruit indehiscent. Juncaceae: leaves clustered at base, culms solid, tepals usually 6 in 2 whorls, flowers bisexual or unisexual, fruit capsular with many seeds. Cyperaceae: leaves clustered at the base, culms triangular, perianth often bristle-like or reduced, flowers bisexual, fruit a nutlet.

Habit culms tufted, grouped along linear creeping rhizomes, tangled together, or bamboo-like (up to 3.5 m tall). **Culms** usually solid, branched or not, usually cylindrical (may be flattened or square) and straight (may be curly), photosynthetic, often glabrous, but simple or fan-like hairs present in some Australian genera. Short sterile culms often present at the nodes. **Leaves** blades absent, only leaf sheaths present, usually brown, dehisce by splitting right down to the base leaving an abscission ring on the culm. Sheaths identical in male and female plants, often topped with a bristle or mucro. **Inflorescences** very variable, usually shortly pedicellate, flowers aggregated into spikelets, subtended by a bract. Spikelets usually numerous, but may be reduced to a single spikelet. **Flowers** unisexual, no more than 1 cm across, perianth of 6 tepals in 2 whorls, usually chaffy, sometimes membranous or very reduced. Male flowers with 1–3 anthers opposite inner tepals, usually exserted at anthesis, pollen copious. Female flowers staminodes 2–3 or absent, 1–3 style branches (consistent within a species), ovary 3-locular. **Fruit** a 3-locular capsule or 1-locular nut. **Seeds** often with surface ornamentation.

Literature: Dorrat-Haaksma & Linder (2012); Linder (1986); Linder *et al.* (1998); Meney & Pate (1999).

51 genera and 572 species. Widely distributed in the Southern Hemisphere, but almost all the species diversity is in the Cape Floristic Province and south-west Australia. Varying circumscriptions depending on whether Anarthriaceae, Centrolepidaceae and Lyginiaceae are included (all are included according to APG IV).

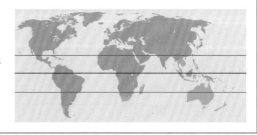

HABIT

Tufted perennials: **1** *Staberoha banksia.* Creeping rhizomes: **2** *Ceratocaryum argenteum.* Bamboo-like: **3** *Elegia capensis.*

STEMS

Sterile growth at nodes: **4** *Rhodocoma capensis.* Leaf sheath split to base: **5** *Elegia mucronata.* Flowers in spikelets: **6** *Baloskion gracile.*

FLOWERS AND FRUIT

Flowers unisexual (female): **7** *Lepidobolus quadratus.* Flowers unisexual (male): **8** *Tremulina tremula.* Fruit: **9** *Ceratocaryum fistulosum.*

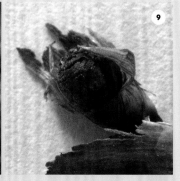

Poaceae

David A Simpson & Martin Xanthos

Stolons and tillers in perennial species
Leaves 2-ranked
Culms rounded or flattened, often, hollow, noded
Lemma and palea present as part of floret
Fruit a caryopsis

Rhizomatous, stoloniferous or annual, sometimes woody **herbs**. **Culms** noded. **Leaves** 2-ranked, ligule present. **Inflorescence** with spikelets in open or contracted panicles, racemes or spikes. **Lemma** and palea present as part of a floret. **Fruit** a caryopsis.

LEFT TO RIGHT:
Polypogon monspeliensis;
Poa annua

Characters of similar families: Cyperaceae: culms 3-sided, leaves usually 3-ranked, flowers minute, perianth reduced to bristles, rarely tepals, or 0, fruit a 1-seeded nutlet. **Juncaceae:** stems rounded and without nodes, flowers mostly hermaphrodite, minute and regular, perianth of 6 tepals, fruit a capsule with 3–many seeds. **Restionaceae:** plants mostly dioecious, leaves usually reduced to sheaths, flowers mostly unisexual, minute and regular, perianth of 6 tepals, fruit a 1–3-locular capsule or nutlet.

Herbs, rhizomatous, stoloniferous or annual, sometimes woody (bamboos). **Culms** rounded or flattened, often hollow, noded. **Leaves** solitary at the nodes, sometimes crowded at the base of the stem, alternate and usually in 2 ranks, consisting of sheath, ligule and blade; sheath with margins free and overlapping or, rarely, more or less fused; ligule at junction of blade and sheath, membranous, a row of hairs, rarely absent; blade usually linear, rarely broad and short, flat, rolled or folded, parallel-veined. **Inflorescence** comprising spikelets arranged in open or contracted panicles, racemes or spikes. **Spikelets** of 1–many flowers (florets), distichous, sessile on a slender axis (the rachilla), subtended by 2 (rarely 1) empty glumes. **Florets** mostly bisexual, sometimes unisexual or sterile, usually with ovary, stamens, and 2–3 minute fleshy scales (lodicules), the whole between 2 bracts that are known as the lemma (the lower bract) and the palea. **Stamens** 1–6 but most commonly 3. **Stigmas** usually 2, plumose. **Fruit** a caryopsis, with the pericarp adhering to the seed.

Literature: Barkworth *et al.* (2003, 2006); Chen *et al.* (2006); Clayton & Renvoize (1986); Kellogg (2015).

About 760 genera and 12,000 species. Worldwide, including Antarctica. Poaceae are dominant components of many non-forest ecosystems. The Poaceae include some of the world's most important food plants, with wheat (*Triticum*), rice (*Oryza*) or maize (*Zea*) forming the basis of most human diets.

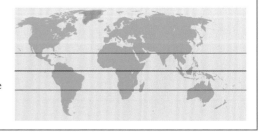

HABIT

Delicate to very robust herbs: **1** *Dactylis glomerata*, **2** *Arundo donax*, **3** *Hordeum jubatum*, **4** *Poa pratensis*, **5** *Cortaderia selloana*, **6** *Aegilops* sp. and **7** *Phyllostachys platyglossa*.

SPIKELETS

8 *Avena eriantha*, **9** *Briza media*, **10** *Bromus danthoniae* and **11** *Lolium temulentum*.

Papavaraceae

Renata Borosova

Sap present
Stipules absent
Leaves simple or compound
Flowers bisexual
Ovary superior

Herbs. Sap milky or watery. **Leaves** simple or compound, usually variously divided, **petioles** clasping. **Flowers** large, nodding in bud; actinomorphic or zygomorphic; **perianth** in 3 whorls, **stamens** numerous. **Fruit** a capsule.

LEFT TO RIGHT:
Papaver somniferum;
Platystemon californicus.

Characters of similar families: Berberidaceae: sap absent, sepals 6–9, petaloid. **Cleomaceae:** stipules usually present, sap absent, 4 petals and 4 sepals, a characteristic smell. **Convolvulaceae, Caryophyllaceae** and **Linaceae:** sap absent, flowers 5-merous. **Gentianaceae:** sap absent, glabrous plants, leaves with entire margins, corolla tubular. **Hypericaceae:** leaf blades with glands. **Limnanthaceae:** sap absent, leaves alternate, inflorescences axillary, sepals 3–5, persistent in fruit. **Ranunculaceae:** sap absent, carpels free.

Mostly herbaceous **annuals** and **perennials**, a few are woody shrubs or climbers, often with white, yellow, orange or watery latex. **Stipules** absent. **Leaves** alternate or spirally arranged, rarely opposite or whorled, sometimes in basal rosettes; simple, pinnately or palmately compound, usually with clasping petioles; margin entire, lobed or deeply dissected. **Inflorescence** usually racemose, cymose or solitary, sometimes a scapose thyrse; terminal or axillary. **Flowers** generally large, bisexual, actinomorphic or zygomorphic, flower buds often nodding; sepals 2–(3), free, sometimes fused and often caducous, petals (2–)4–6, free, showy, sometimes with a sack or spur, rarely absent or more (–12); stamens usually numerous in several whorls, sometimes 2, 4, 6 or 8, filaments sometimes petaloid, free or fused, sometimes with nectaries, anthers opening by slits; ovary superior, composed of 2–many carpels, usually fused. **Fruit** usually a capsule, sometimes a nut or loment. **Seeds** oily, sometimes arillate.

Literature: Christenhusz *et al.* (2017); Culham (2007b); Kadereit (1993); Plants of the World Online (2019); World Checklist of Vascular Plants (2021); Zhang *et al.* (2008).

42 genera and 1,029 species. Papaveraceae includes Fumariaceae and Pteridophyllaceae. Widespread throughout northern temperate regions, with a few genera in Central and South America, and eastern and southern Africa. Used in horticulture, as a condiment (*Papaver somniferum*) and medicinally (*P. somniferum* and *P. bracteatum*) because of the high content of morphinane alkaloids.

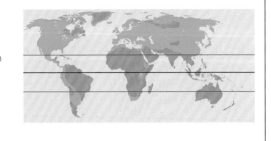

HABIT

Herbs: **1** *Papaver tianschanicum*. Shrubs: **2** *Romneya coulteri*. Climbers: **3** *Dactylicapnos scandens*.

FLOWERS

Zygomorphic: **4** *Corydalis pumila*, **5** *Fumaria densiflora* and **6** *Lamprocapnos spectabilis*. Actinomorphic: **7** *Eschscholzia californica*, **8** *Argemone subfusiformis* and **9** *Chelidonium majus*.

FRUIT

Capsule: **10** *Papaver somniferum*, **11** *Corydalis fedtschenkoana* and **12** *Dactylicapnos scandens*.

Berberidaceae

Renata Borosova

Stipules usually absent
Leaves simple or compound
Leaves spiral or opposite
Flowers bisexual
Ovary superior

Stems with or without spines. **Perianth** free, multiseriate; **sepals** 6–9, petal-like in 2–3 whorls; **petals** 6, hooded, pouched or spurred; **stamens** 6; **anthers** open by valves or slits.

LEFT TO RIGHT:
Berberis thunbergii;
Jeffersonia diphylla.

Characters of similar families: Cistaceae: leaves usually opposite, flower parts 3 or 5. Clethraceae: flowers 5-merous, sepals sometimes fused. Papaveraceae: often with white, yellow or watery latex, 2 sepals. Rhamnaceae: flowers 4- or 5-merous with a hypanthium or calyx tube.

Woody **shrubs** or perennial **herbs** with rhizomes or tubers. **Stipules** usually absent. **Leaves** spiral, rarely opposite, sometimes evergreen and persistent, simple to pinnately or ternately compound and in *Berberis* often transformed into spines. **Inflorescences** terminal or axillary, in racemes, spikes, umbels, cymes, panicles or flowers solitary. **Flowers** bisexual and actinomorphic; perianth free, whorled, sometimes absent (*Achlys*); sepals 6–9, petaloid in 2–3 whorls; petals usually 6 and bearing nectaries or reduced to nectariferous sacs or scales; stamens mostly in 2 whorls, opposite petals, usually 6, rarely 4 (*Epimedium*) or 9–18 (*Podophyllum*); anthers open by lengthwise slits (*Nandina* and *Podophyllum*) or by valves hinged at the top; ovary superior with a single carpel, style terminal, often persistent, rarely stigma sessile. **Fruit** a berry (*Nandina*, *Berberis* and *Podophyllum*), capsule (*Bongardia*, *Leontice*, *Epimedium* and *Vancouveria*), follicle or utricle with 1–many seeds.

Literature: Christenhusz *et al.* (2017); Culham (2007a); Loconte (1993); Plants of the World Online (2019); World Checklist of Vascular Plants (2021); Yang *et al.* (2011). George (1997).

13 genera and 769 species. Worldwide in temperate regions and on tropical mountains. Used in horticulture, some species have fruits that are high in vitamin C and that are used in cooking or made into preserves.

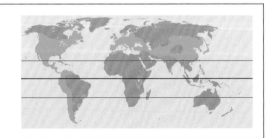

HABIT, LEAVES AND SPINES

Shrubs: **1** *Berberis amurensis*. Herbs: **2** *Podophyllum cymosum*. Leaves ternate: **3** *Achlys triphylla*. Leaves pinnate: **4** *Bongardia chrysogonum*. Leaves with spines: **5** *Berberis trifolia*.

INFLORESCENCES AND FLOWERS

6 *Caulophyllum thalictroides*, **7** *Epimedium leptorrhizum*, **8** *Gymnospermium albertii*, **9** *Nandina domestica* and **10** *Podophyllum peltatum*.

FRUIT

Berry: **11** *Berberis mucrifolia* and **12** *Berberis aquifolium*. Capsule: **13** *Bongardia chrysogonum* and **14** *Leontice leontopetalum*.

Ranunculaceae

Renata Borosova

Leaves simple or compound
Leaves usually alternate
Flowers bisexual
Perianth parts free
Ovary superior

Herbs, **shrubs** or **climbers**. **Stipules** absent or minute. **Leaves** simple, trifoliolate, palmate or 1–2-pinnate. **Flower** parts many, free and spirally arranged, **stamens** numerous. **Fruit** usually a head of achenes or follicles.

LEFT TO RIGHT:
Anemonopsis macrophylla;
Thalictrum diffusiflorum.

Characters of similar families: Capparaceae: trees, shrubs or lianas, sometimes spiny; leaf margin entire; 4 sepals and 4 petals. **Cleomaceae:** stipules usually present, 4 petals and 4 sepals. **Papaveraceae:** often with white, yellow or watery latex, 2 sepals, carpels usually fused. **Rosaceae:** stipules usually present, flowers with a hypanthium. **Salicaceae:** leaves simple with salicoid teeth, unisexual.

Perennial, annual, sometimes aquatic **herbs**, **shrubs** or woody **climbers** (*Clematis*). **Stipules** absent or minute. **Leaves** alternate, rarely opposite or whorled, simple, trifoliolate or palmately, pedately or 1–2-pinnately compound and usually petiolate; margin usually lobed or toothed. **Inflorescences** terminal or axillary, racemose (*Clematis*), cymose, paniculate or flower solitary. **Flowers** usually bisexual and actinomorphic, sometimes zygomorphic, flower parts usually many, free, can be clawed, spurred or hooded, spirally arranged; sepals petaloid or sepaloid, (3–)5–8 or more, often caducous; petals (2–)5–many, sometimes absent, free, usually with a nectary gland; stamens usually numerous, free, arranged in spirals or whorls, anthers usually adnate and dehiscing laterally; carpels numerous to few or rarely 1, usually free; ovary superior, with a solitary pendulous or erect ovule or one to few ovules. **Fruits** achenes, follicles, berries or rarely capsules (*Nigella*), often aggregated into heads, sometimes with elongated plumes that are formed by the persistent styles (*Clematis*). **Seeds** small, endosperm oily, abundant.

Literature: Christenhusz *et al.* (2017); Culham (2007d); Fu & Zhu (2001); Plants of the World Online (2019); Tamura (1993); World Checklist of Vascular Plants (2021).

52 genera and 3,803 species. Worldwide except for Antarctica and desert regions of Africa and Australia. Ranunculaceae are most common in temperate and cold regions of the northern hemisphere; in the tropics, they are less common and then usually at high altitudes. Commonly used in horticulture, many are toxic and medicinal plants, with very few edibles (*Nigella sativa*).

HABIT
Herbs: **1** *Caltha palustris* and **2** *Delphinium dasyanthum*. Climbers: **3** *Clematis montana* var. *longipes*.

FLOWERS
Flowers actinomorphic: **4** *Helleborus orientalis*, **5** *Thalictrum isopyroides*, **6** *Anemone nemorosa* and **7** *Trollius yunnanensis*.
Flowers zygomorphic: **8** *Aconitum hemsleyanum* and **9** *Delphinium sclerocladum*. Terminal inflorescence: **10** *Actaea simplex*.
Dissected bracts: **11** *Nigella damascena*.

FRUIT
Hooked achenes: **12** *Ranunculus arvensis*. Plumose achenes: **13** *Clematis integrifolia*. Berry: **14** *Actaea rubra*. Capsule: **15** *Nigella damascena*.

75

Proteaceae

Richard Wilford

Stipules absent
Leaves simple or compound
Leaves alternate or opposite
Flowers usually bisexual
Ovary superior

Trees and **shrubs** often with clusters of short lateral roots. **Leaves** simple or compound, usually alternate, coriaceous. **Inflorescence** a panicle, raceme or condensed head. **Perianth** of 4 petaloid tepals. **Fruit** a follicle, achene, nut or drupe.

LEFT TO RIGHT:
Grevillea hookeriana: deeply lobed leaves with narrow segments; *Banksia littoralis* with inflorescence and fruit.

Characters of similar families: Myrtaceae: trees or shrubs with peeling bark; flowers with separate calyx and corolla; stamens numerous, rarely 10 or fewer; ovary usually inferior, rarely semi-inferior; fruits capsular or fleshy.

Trees and **shrubs**, creeping to upright, bisexual or sometimes dioecious, often with clusters of short lateral roots (proteoid roots). **Stipules** absent. **Leaves** alternate, rarely opposite or verticillate, usually coriaceous, petiolate or sessile, simple or pinnately to bipinnately or rarely palmately compound, entire to lobed or pinnate, often toothed, venation usually pinnate. **Inflorescence** simple or compound, axillary or often terminal; racemes, panicles or condensed heads, or cones with solitary flowers; sometimes surrounded by involucral bracts. **Flowers** usually bisexual, actinomorphic or zygomorphic, usually 1 or 2 in axils of bracts. **Perianth** of 4 petaloid tepals, free or variously fused. **Stamens** 4, opposite tepals, filaments partly or wholly adnate to tepals or rarely free; anthers basifixed. **Ovary** superior, sessile or stipitate, carpels 1, style sometimes very long and exceeding the perianth, 1–4 scale-like or fleshy hypogynous glands usually present around ovary. **Fruit** dehiscent or indehiscent, a woody or coriaceous follicle, achene, nut or drupe; seeds 1–many, sometimes winged.

Literature: Christenhusz *et al.* (2017); Manning (2007); Weston (2007).

A predominantly southern hemisphere family with 38 out of 80 genera in the temperate zone and about 1,750 species in total. Centres of diversity are in Australia and southern Africa. Some species are important in the cut flower industry, including *Protea*, *Leucospermum* and *Banksia*.

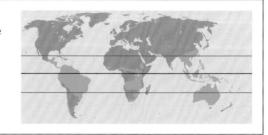

HABIT

Trees or shrubs: **1** *Hakea lissosperma*. Shrub with terminal inflorescences and flowers with long, curved styles: **2** *Leucospermum cordifolium* 'Yellow Bird'.

LEAVES

Leaves with a serrated margin: **3** *Banksia serrata*. Leaves felted and with 3–10 apical teeth: **4** *Leucospermum conocarpodendron*. Leaves entire or with few teeth towards tip: **5** *Banksia integrifolia*.

INFLORESCENCE AND FLOWERS

Dense cylindrical head: **6** *Banksia integrifolia*. Terminal head with coloured bracts: **7** *Protea cynaroides*. Flowers in spikes at branch tips: **8** *Mimetes cucullatus*.

FRUIT

Follicles: **9** *Grevillea banksii*. Cone formed from overlapping woody bracts: **10** *Leucadendron* 'Safari Sunset'.

Buxaceae

Harry Baldwin & Tony Kirkham

Stipules absent
Leaves simple
Leaves alternate
Flowers actinomorphic
Ovary superior

Trees, **shrubs** or **herbs**, evergreen. **Stipules** absent. **Leaves** alternate. **Flowers** actinomorphic and unisexual, styles persistent in fruit. **Fruit** capsules or drupes, dehiscing loculicidally into 3 spreading, 2-horned valves, seeds black and shiny.

LEFT TO RIGHT:
Buxus sempervirens;
Buxus henryi: type specimen.

Characters of similar families: Asparagaceae, *Ruscus*: dioecious, leaves reduced to scales replaced by cladodes, flowers arising from centre of cladode; fruits fleshy, red. **Asparagaceae**, *Danae*: monoecious, leaves reduced to scales replaced by lanceolate cladodes, flowers terminal raceme, fruits fleshy, red.

Evergreen **trees**, **shrubs** or rhizomatous perennial **herbs**, monoecious or rarely dioecious. **Stipules** absent. **Leaves** alternate (*Sarcococca*), opposite or sometimes decussate (*Buxus*), simple, petiolate, rarely sessile, entire, rarely dentate, blades with bases decurrent as internodal folds on the branches (*Buxus*), pinnately veined, less often tripliveined. **Inflorescence** axillary or terminal botryoids or spikes, flowers subtended by decurrent bracts, the female with prophylls. **Flowers** actinomorphic, unisexual; staminate flowers with 4 inconspicuous bract-like tepals; petals absent; male flowers with usually 4 stamens opposite the tepals, inserted on angular bracts, often inserted around a pistillode, nectariferous; female flowers with 4–6 tepals, bract-like; ovary syncarpous, superior, with free stylodia, subulate, rarely connate at base, persistent in fruit (*Buxus*), stigmatic area decurrent along the ventral fold, (2–)3(–4) carpellate, ovules usually 2 per locule, anatropous, bitegmic. **Fruit** a capsule or drupe-like, with persistent stylodia; capsules dehiscing loculicidally into 3 spreading, 2-horned valves, partly enclosing and ejecting the seeds. **Seeds** usually (2–)4–6 black or blue, shining, caruncle present (*Buxus*); endosperm copious, fleshy, oily, the cotyledons thin, flat.

Literature: Batdorf (2005); Köhler (2009).

Buxaceae comprises 6 genera with ca. 140 species. Commonly cultivated temperate genera include *Buxus*, *Sarcococca* and *Pachysandra*, whereas *Styloceras*, *Haptanthus* and *Didymeles* are generally of tropical origin. Buxaceae has no close relatives; the nearest branch on the phylogenetic tree of angiosperms is *Trochodendron* and *Tetracentron*, which were not considered before the molecular era.

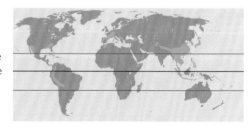

HABIT AND LEAVES

Evergreen tree: **1** *Buxus sempervirens*. Shrub: **2** *Sarcococca hookeriana*. Rhizomatous perennial: **3** *Pachysandra terminalis*. Elliptic, opposite, evergreen leaves: **4** *Buxus balearica*.

INFLORESCENCES

Erect terminal inflorescence: **5** *Pachysandra terminalis*. Axillary inflorescence, tepals cup-shaped, 4 stamens: **6** *Buxus balearica*.

FLOWERS AND FRUIT

Male flower with 4 tepals and 4 stamens: **7** *Sarcococca hookeriana* var. *digyna* 'Tony Schilling'. Stigma held in leaf axil, base of rachis: **8** *Pachysandra terminalis*. Fruit a capsule: **9** *Buxus sempervirens*.

79

Hamamelidaceae

Harry Baldwin & Tony Kirkham

Stipules present
Leaves simple
Leaves alternate or opposite
Flowers actinomorphic
Ovary inferior or superior

Trees or **shrubs**. **Leaves** with stellate indumentum; stipules borne on stem, usually paired. **Inflorescence** a spike or head, usually compact; perianth reduced or absent. **Fruit** a woody capsule, dehiscent; seed dispersal ballistic.

LEFT TO RIGHT:
Hamamelis virginiana;
Disanthus cercidifolius: type specimen.

Characters of similar families: Altingiaceae: trees; monoecious; stipules usually present, borne on petiole; leaf blade palmately lobed (entire in subtropical species); flowers unisexual; sepals and petals absent; styles persistent in fruit; infructescences globose, styles often indurate and spiny in fruit. **Cercidiphyllaceae**: trees; dioecious; leaves opposite, oval; flowers in terminal heads, 16–35 stamens; perianth absent; stamens red; fruit a small follicle.

Habit, trees or shrubs, deciduous, or evergreen. **Buds** perulate or naked. **Stipules** small, deciduous, borne on stem, usually paired (solitary in *Mytilaria*). **Leaves** simple, petiolate, rarely spiral, opposite, sometimes subopposite, simple, tricuspidate or lobed; venation camptodromous or actinodromous and brochidodromous or craspedodromous; margin entire or dentate; indumentum stellate or tufted hairs. **Inflorescences** spikes, heads, racemes or (condensed) thyrses or panicles. **Flowers** actinomorphic, rarely zygomorphic (*Rhodoleia*); mostly yellow, white, greenish or red; bisexual, andromonoecious or unisexual; perianth reduced or absent; sepals absent or 4–5–(10), imbricate or persistent; petals absent or 4 or 5, often ribbon-like and circinate in bud; stamens (1)–2–5 or numerous, alternipetalous; anthers basifixed, dehiscing by valves or by a longitudinal slit; ovary 1(–2), inferior to superior, carpels (1)–2, often free at apex, 2(–3) locules, styles 2, stigmas 2 sometimes decurrent. **Fruit** dehiscent capsules, endocarp woody or leathery. **Seeds** 1 or more per carpel, dispersed by ballistic ejection (other than *Rhodoleia*), mostly hard and black.

Literature: Angiosperm Phylogeny Group (2003); Bittrich *et al.* (1993); Li & Bogle (2001); Simpson (2019).

Hamamelidaceae contains 26 genera (8 being monotypic) and ca. 100 species. The family has a wide but scattered and relict distribution, with the greatest diversity in E Asia. *Hamamelis*, *Disanthus* and *Parrotia* are used extensively in ornamental horticulture. Tropical groups have persistent leaves with mainly brochidodromous venation, whereas temperate genera mostly have deciduous leaves with craspedodromous venation.

HABIT

Multi-stemmed deciduous tree: **1** *Parrotia persica*. Winter-flowering shrub: **2** *Hamamelis* x *intermedia*.

INFLORESCENCES

Flowers emerging: **3** *Parrotia persica*. Inflorescences in cylindrical terminal spikes with protruding anthers: **4** *Fothergilla major*. Flowers borne on drooping racemes: **5** *Corylopsis glabrescens*.

FLOWERS

Uncurling ribbon-shaped petals: **6** *Hamamelis mollis*. Ovate petaloid whitish bracts: **7** *Parrotiopsis jacquemontiana*. Over mature inflorescence: **8** *Parrotiopsis jacquemontiana*.

FRUIT

Ovules beginning to swell: **9** *Fothergilla major*. Capsules woody: **10** *Corylopsis glandulifera*.

Saxifragaceae

Richard Wilford

Leaves simple or compound
Leaves alternate
Flowers bisexual
Flowers usually actinomorphic
Perianth parts free

Perennial **herbs**, rarely annual or biennial. **Leaves** alternate, in a basal rosette. **Inflorescence** terminal, cymous to racemous, rarely solitary. **Petals** usually 5 or absent. Stamens 5 or 10, **Fruit** a capsule.

LEFT TO RIGHT:
Saxifraga marginata;
Heuchera cylindrica.

Characters of similar families: Crassulaceae: succulent stems and leaves, some shrubs, stipules absent, ovary superior or partly inferior, fruit a follicle. **Ranunculaceae:** stipules absent or minute, stamens numerous and arranged in whorls, fruit an achene, follicle or berry.

Perennial **herbs**, rarely annual or biennial, often rhizomatous. **Stipules** usually present or leaf bases sheathing. **Leaves** usually in a basal rosette, sometimes cauline, usually alternate, simple or compound, rarely peltate (*Darmera*), venation pinnate or sometimes palmate, margins entire, lobed, crenate or toothed, sometimes with lime-secreting hydathodes. **Inflorescence** terminal, bracteate, cymose to racemose, occasionally solitary flowers. **Flowers** usually actinomorphic, sometimes zygomorphic, usually bisexual, homostylous (heterostylous in *Jepsonia*), hypanthium free or partly fused with the ovary. **Calyx** lobes (3–)5(–10), fused with the hypanthium. **Petals** (4–)5(–6) or absent, clawed, rarely cleft or dissected. **Stamens** usually 5 or 10, free, anthers basifixed. **Ovary** inferior or semi-inferior, carpels 2 or rarely 3, each carpel topped with a stylodium and capitate stigma. **Fruit** a capsule, seeds generally small and numerous, rarely winged.

Literature: Christenhusz *et al.* (2017); Soltis (2007); Well & Elvander (2009).

35 genera and ca. 640 species. Widespread in temperate, arctic and alpine regions of the northern hemisphere and S America. Historically, Saxifragaceae contained genera that are now segregated into other families, including Hydrangeaceae and Grossulariaceae. Genera such as *Saxifraga*, *Heuchera* and *Astilbe* contain popular ornamental plants.

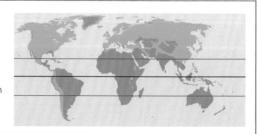

HABIT AND LEAVES

Perennial herbs with a basal rosette of leaves: **1** *Saxifraga cotyledon*. **2** *Bergenia stracheyi*. Leaves palmately compound: **3** *Rodgersia aesculifolia*. Simple leaves with toothed margins: **4** *Saxifraga spathularis*.

INFLORESCENCES

Terminal cymous inflorescence: **5** *Darmera peltata*. **6** *Mukdenia rossii*. Many-flowered raceme: **7** *Tiarella cordifolia*. Solitary flowers: **8** *Saxifraga burseriana*.

FLOWERS AND FRUIT

Actinomorphic flowers with 5 petals: **9** *Saxifraga cebennensis*. Clawed petals, 2 carpels: **10** *Bergenia emeiensis*. Zygomorphic flowers: **11** *Saxifraga stolonifera*. Fruit a capsule: **12** *Saxifraga canaliculata*.

Crassulaceae

Richard Wilford

Stipules absent
Leaves simple
Leaves alternate or opposite
Flowers actinomorphic
Flowers bisexual

Succulent perennials or **shrubs**, rarely annuals or biennials. **Inflorescence** terminal or axillary, with actinomorphic flowers. Sepals and petals of equal number. **Carpels** as many as petals. Small seeds held in follicles.

LEFT TO RIGHT:
Sempervivum arachnoideum;
Dudleya stolonifera showing
2-branched cymes.

Characters of similar families: Saxifragaceae: leaves may be leathery but rarely succulent, leaves usually in a basal rosette, ovary inferior or semi-inferior, carpels 2 or rarely 3, fruit a capsule. Aizoaceae: leaf blade flat, terete, or triquetrous (3-angled), occasionally scale-like. Petals linear or absent, 0–250(–300). Stamens 1–500(–700), fruits usually capsules, sometimes indehiscent berries or nut-like.

Succulent perennial herbs, rarely annuals or biennials, some **shrubs**. **Stipules** absent. **Leaves** opposite or alternate, rarely verticillate, sometimes in a basal rosette, simple (compound in *Kalanchoe*), margins usually entire sometimes toothed or lobed. **Inflorescence** terminal or axillary, usually cymous, sometimes a raceme or spike, occasionally flowers solitary. **Flowers** bisexual (sometimes unisexual in *Rhodiola*), actinomorphic, perianth and androecium hypogynous or weakly perigynous, nectaries as scales at base of carpels. **Calyx** (3–)4–5(–20) sepals, free or united into tube, entire. **Petals** same number as sepals, free or united into tube. **Stamens** as many or twice as many as petals, free or fused with corolla tube; anthers basifixed and dehiscing longitudinally. **Carpels** superior or partly inferior, free or united at base, as many as petals. **Fruit** usually whorls of follicles, membranous or coriaceous; seeds 1–20+ per carpel, small, brownish.

Literature: Christenhusz *et al.* (2017); Kunjun *et al.* (2001); Moran (2009).

36 genera (29 in temperate regions) and ca. 1,400 species, although there are only 7 genera if *Sedum* is expanded to include all of the genera in subfamily Sempervivoideae. Widely distributed across the northern and southern hemispheres. Some genera have medicinal uses and several, including *Sedum* and *Sempervivum*, are grown as ornamentals.

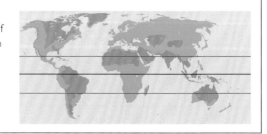

HABIT

Perennial rosettes: **1** *Sempervivum tectorum*. Growing in a stone wall: **2** *Umbilicus rupestris*. Trailing stems with evergreen leaves: **3** *Sedum kamtschaticum*.

LEAVES

Leaves succulent, semi-terete, mucronate: **4** *Petrosedum sediforme*. Succulent, lobed leaves: **5** *Rhodiola pachyclados*. Farinose leaf surface: **6** *Dudleya farinosa*.

FLOWERS AND FRUIT

Actinomorphic flowers: **7** *Sedum album*. Cymous inflorescence: **8** *Hylotelephium* 'Matrona'. Fruit a follicle: **9** *Rosularia sempervivum*.

Vitaceae

Anna Trias-Blasi

Stipules at petiole base	
Leaves simple or compound	
Leaves alternate	
Flowers actinomorphic	
Ovary superior	

Climbing by leaf-opposed tendrils. **Leaves** simple, lobed or compound. **Inflorescence** leaf-opposed, axillary or pseudoterminal. **Flowers** actinomorphic, generally with intrastaminal disk. Petals valvate and cucullate. **Fruit** a berry.

LEFT TO RIGHT:

Parthenocissus quinquefolia: note leaf-opposed tendril;

Vitis bryoniifolia: note the leaf-opposed tendril.

Characters of similar families: Cucurbitaceae: stipules absent, tendrils at 90° of the petiole, inferior ovary. **Convolvulaceae**: tendrils absent, large showy corolla. **Smilacaceae**: tendrils in pairs not leaf-opposed, leaf margin entire.

Usually **climbers** with leaf-opposed **tendrils**, simple or compound. **Stipules** small, caducuous, sometimes persistent. **Leaves** alternate, simple or palmately, pedately or pinnately compound; margins toothed to lobed. **Inflorescence** paniculate, corymbose or thyrsoid, leaf-opposed, pseudo-terminal or axillary. **Flowers** small, actinomorphic, generally pedicellate, sometimes sessile or subsessile, with bracts and bracteoles, actinomorphic, hypogynous, bisexual or unisexual, (3–)4–5(–6)-merous; sepals (3–)4–5(–6), generally cupulate; petals free or distally connate forming a calyptra, valvate (3–)4–5(–6), apex generally cucullate; stamens (3–)4–5(–6), antepetalous, anthers with longitudinal dehiscence, generally tetrasporangiate, often with intrastaminal disk, cupular, annulate or with 4 separate glands, mostly adnate to ovary, sometimes absent; ovary superior with 2 carpels, 2 locules, 2 ovules per locule; style simple, stigma mostly inconspicuous, sometimes capitate, discoid or 4-lobed (*Tetrastigma* sp.). **Fruit** a berry, 1–4-seeded. **Seeds** endotestal with an abaxial chalazal knot and an adaxial raphe; endosperm oily, proteinaceous, ruminate.

Literature: Schieber *et al.* (2001); Trias-Blasi *et al.* (2020); Wen *et al.* (2018).

16 genera and ca. 950 species with a pantropical and (warm) temperate distribution. Genera that are mainly or completely temperate are *Ampelopsis*, *Nekemias*, *Parthenocissus*, *Vitis* and *Yua*. APG IV recognises two subfamilies: Vitioideae and Leeoideae, while others treat them as two separate families. Grapes (*Vitis* sp.) are the world's largest fruit crop.

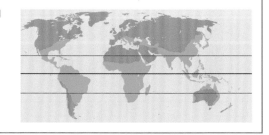

LEAVES AND TENDRILS

Simple leaves: **1** *Ampelopsis glandulosa* var. *brevipedunculata*. Lobed leaves: **2** *Parthenocissus tricuspidata*. Compound leaves: **3** *Parthenocissus quinquefolia*. Leaf-opposed tendril: **4** *Rhoicissus* sp.

FLOWERS

5-merous flowers: **5** *Ampelopsis aconitifolia*. 4-merous flowers: **6** *Cyphostemma omburense*. Calyptra: **7** *Vitis giradiana*. Stamens, stigma, ovary and disk: **8** *V. giradiana*.

FRUIT

Berry: **9** *Vitis vinifera* and **10** *Ampelopsis glandulosa* var. *brevipedunculata*.

Leguminosae–Papilionoideae

Ruth Clark

Stipules present
Leaves usually 1-pinnate
Flowers zygomorphic
Sepals united at base
Ovary superior

Leaves 1-pinnate (many 3-foliolate), 1-foliolate or simple, never bipinnate. '**Pea flower**', 5 petals (1 standard petal, 2 wings, 2 keels). **Stamens** usually (9–)10(–many), often fused. **Seeds** without pleurogram.

LEFT TO RIGHT:
Wisteriopsis japonica;
Desmodium canescens.

Characters of similar families: **Polygalaceae:** leaves simple, no stipule, usually 3 petals, 2 of the 5 sepals united, or enlarged and petaloid, seeds often with hairs. **Caesalpinioideae s.l.** (excluding the mimosoid clade): median petal innermost, sepals generally free. **Moringaceae:** 5 stamens and 5 staminodes, joined at the base into a disk.

Herbs, shrubs, trees, lianas, or **twiners**. Root nodules generally present. **Stipules** generally present. **Leaves** usually alternate, occasionally opposite; usually 1-pinnate (many 3-foliolate), occasionally 1-foliolate or simple, sometimes with tendrils, never bipinnate, with pulvina; lamina margin usually entire, sometimes toothed. **Inflorescence** with flowers solitary or in racemes, pseudo-racemes, spikes, or panicles, axillary or terminal. **Flowers** strongly zygomorphic; calyx 5-lobed, the 2 upper lobes sometimes fused, the sepals united at the base to form a tube; petals usually 5 (1 standard petal (vexillum, banner, median), 2 wing petals, and 2 keel petals (often fused)), standard petal overlapping the others, petals imbricate in bud; stamens (9–)10–many, often fused, sometimes dimorphic; anthers dehiscing by longitudinal slit; ovary superior, unilocular (rarely bilocular). **Fruit** most often a typical pod, but may be a drupe, a loment, or a samara. **Seeds** (if hard) with complex hilar valve; pleurogram absent. Pollen in single grains.

Literature: Legume Phylogeny Working Group (2017); Lewis *et al.* (2005); Various contributors in Advances in Legume Systematics (1978–ongoing).

An almost cosmopolitan subfamily of 531 genera and ca. 14,000 species. Found in dry and wet environments, lowland and montane habitats. Species in this subfamily have many uses as foods, medicines, forage, timber, agroforestry products, soil improvers, insecticides, fish poisons, dyes and resins.

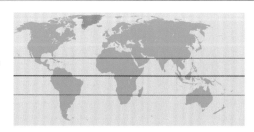

VEGETATIVE CHARACTERS

3-foliolate leaf with stipule: **1** *Lotus corniculatus*. 1-pinnate leaf: **2** *Tephrosia* sp. Stipule: **3** *Trifolium stellatum*.

FLOWERS AND INFLORESCENCES

Papilionate flower: **4** *Lathyrus cicero*, **5** *Medicago sativa*, **6** *Clianthus puniceus* and **7** *Lotus maritimus*. Fused calyx, free stamens: **8** *Sophora* aff. *tetraptera*. Racemes: **9** *Wisteria* sp., **10** *Lupinus polyphyllus* and **11** *Astragalus hamosus*. Head of small flowers: **12** *Trifolium pratense*. Fused stamens: **13** *Ulex europaeus*.

FRUITS AND SEEDS

Typical pod: **14** *Vicia hirsuta* and **15** *Calicotome villosa*. Lomentaceous fruit: **16** *Arthroclianthus* sp. Small, inflated pod: **17** *Onobrychis saxatilis*. Coiled, spiny fruit: **18** *Medicago polymorpha*. Seeds with hilum and lens: **19** *Phaseolus vulgaris*.

Polygalaceae

Ruth Clark

Stipules absent
Leaves simple
Flowers usually zygomorphic
Flowers bisexual
Ovary superior

Stipules absent. **Leaves** simple, usually alternate, margins entire. **Flowers** usually zygomorphic, resembling a 'pea flower'; stamens usually 8; ovary superior with a single ovule per locule.

LEFT TO RIGHT:
Polygala major;
Polygala nicaeensis subsp. *mediterranea.*

Characters of similar families: Leguminosae (Papilionoideae): stipules present, leaves usually compound, corolla 5-merous, several ovules per locule, anthers dehisce by longitudinal slits. **Malphigiaceae**: stipules present, leaves opposite, winged fruit made up of three samaras.

Herbs, **climbers**, **shrubs** and **trees**, sometimes with swollen nodes. Indumentum when present simple hairs. **Stipules** absent, extra-floral nectaries sometimes present at junction of petiole and stem. **Leaves** simple, spirally arranged, sometimes alternate or whorled, sometimes reduced or absent, petiolate or sessile; lamina margin entire. **Inflorescences** terminal or axillary, usually unbranched, sometimes thyrsoid or in fascicles. **Flowers** bisexual, zygomorphic, each subtended by a bract and 2 bracteoles; sepals 5, quincuncial, free or variously fused, usually with the lowest 2 lobes fused or the 2 lateral lobes expanded and petaloid; petals 3, sometimes 5, free or variously fused, lower petal commonly keel-shaped, pouched, lobed or with a fringed crest; stamens usually 8, filaments often fused, often adnate to petal base, anthers basifixed, 1- or 2-locular, usually dehiscing by an apical pore; intra-staminal disk sometimes present; ovary superior, usually sessile, rarely stipitate, usually 2-locular and with 1 ovule per locule, placentation usually apical. **Fruit** a berry, capsule, samara or drupe. **Seeds** usually arillate, sometimes winged or with hairs.

Literature: Chen *et al. (2008)*; Eriksen & Persson (2007); Pastore (2018); Pastore *et al.* (2017).

A family of 27 genera and ca. 1,200 species, distributed worldwide throughout temperate and tropical areas. In temperate areas, the family is represented primarily by the genus *Polygala*, which has ca. 580 species. Some species are used for horticulture and several have local medicinal uses.

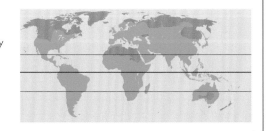

HABIT

Erect, herbaceous habit: **1** *Polygala sanguinea* and **2** *Polygala rugelii*. Simple, alternate leaves: **3** *Polygala amara*. Trailing habit: **4** *Comesperma volubile*. Shrubby habit: **5** *Comesperma ericinum* and **6** *Polygaloides chamaebuxus*.

FLOWERS

Two sepals enlarged and petaloid, lower petal fringed or lobed: **7** *Polygala rupestris*, **8** *Polygala paucifolia*, **9** *Polygala cruciata*, **10** *Polygala vulgaris*, **11** *Polygaloides chamaebuxus*, **12** *Comesperma volubile* and **13** *Polygala lutea*.

FRUIT AND SEEDS

Bilobed fruit: **14** *Polygala vulgaris* and **15** *Polygaloides chamaebuxus*. Seeds with hairs and aril: **16** *Polygala vulgaris*.

Rosaceae

Jo Osborne

Stipules present
Leaves alternate
Flowers actinomorphic
Perianth parts free
Ovary superior/inferior

Trees, **shrubs** and **herbs** (mostly perennial). **Leaves** alternate with stipules. **Flowers** often showy, perianth free, usually of 5 petals and 5 sepals. **Hypanthium** present. **Ovary** superior to inferior.

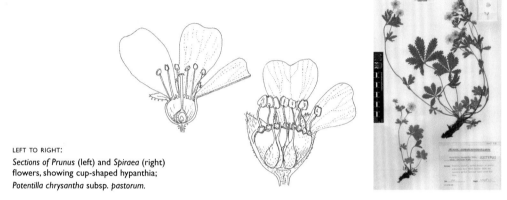

LEFT TO RIGHT:
Sections of Prunus (left) and *Spiraea* (right) flowers, showing cup-shaped hypanthia; *Potentilla chrysantha* subsp. *pastorum*.

Characters of similar families: Celastraceae: stipules absent or small and caducous, flowers small and lacking a hypanthium, disk present. **Malvaceae–Malvoideae:** filaments fused into a tube, flowers lacking a hypanthium. **Ranunculaceae:** stipules usually absent, flowers lacking a hypanthium. **Rhamnaceae:** stipules small, caducous or adapted into spines, floral disk present, calyx tubular at base.

Trees, **shrubs** and **herbs** (mostly perennial), occasionally climbers, sometimes with thorns. **Stipules** usually present, sometimes caducous. **Leaves** usually alternate and spirally arranged, simple or compound, often serrate. **Hairs** usually simple, sometimes glandular, occasionally stellate, sometimes prickles (e.g. *Rosa*). **Inflorescences** diverse, usually compound racemes, often cymes, panicles, fascicles, spikes or heads, occasionally solitary flowers. **Flowers** often large and showy, usually bisexual, occasionally unisexual on dioecious or monoecious plants, actinomorphic, usually with a hypanthium. **Hypanthium** flat to cup-shaped or urceolate. **Sepals** usually 5, free, epicalyx sometimes present. **Petals** usually 5 (rarely absent, e.g. *Alchemilla*), free. **Stamens** few to numerous (often 10–20), filaments free. **Ovary** superior to inferior, of 1–many free or connate carpels, ovules 1–several per carpel. **Fruits** diverse comprising achenes (e.g. *Geum*), follicles (e.g. *Spiraea*), drupes (e.g. *Prunus* (plum)), pomes (e.g. *Malus*), nuculania (e.g. *Prunus* (almond)), or compound fruits:, such as a drupetum (e.g. *Rubus*), glandetum (e.g. *Fragaria*) or pometum (e.g. *Rosa*).

Literature: Gu *et al.* (2003); Judd *et al.* (2008); Kalkman (2004); Phipps (2015); Stevens (2001 onwards).

A cosmopolitan family of 92 genera and ca. 2,805 species, especially diverse in the northern hemisphere. The estimated number of species varies widely according to different authors because several genera in Rosaceae (e.g. *Rubus*) reproduce apomictically, and the apomictic lines can be considered as microspecies. Rosaceae includes many economically important temperate fruits and ornamental plants.

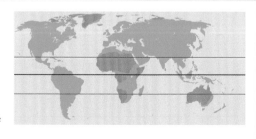

HABIT
Trees, shrubs and herbs: **1** *Prunus armeniaca*, **2** *Prunus petunnikowii*, **3** *Spenceria ramalana* and **4** *Rubus potentilloides*.

FLOWERS AND INFLORESCENCES
Showy flowers with 5 petals and 5 sepals: **5** *Rosa glauca* and **6** *Sibbaldianthe bifurca*. Hypanthium or 'floral cup' at base of petals and sepals: **7** *Neillia affinis* and **8** *Malus ioensis*, flower in section.

FRUITS
Diverse fruits: achenes **9** *Geum aleppicum*, follicles **10** *Spiraea hypericifolia*, drupes **11** *Prunus mahaleb*, pomes **12** *Sorbus microphylla*, drupetum **13** *Rubus fockeanus* or glandetum **14** *Fragaria nilgerrensis*.

Rhamnaceae

Daniel Cahen

Stipules present
Leaves simple
Leaves alternate
Flowers actinomorphic
Perianth parts free

Shrubs (sometimes trees or climbers), often armed. **Stipules** usually present. **Leaves** simple, usually alternate. **Flowers** small, sepals keeled adaxially, petals usually smaller than sepals, hooded and enclosing stamens, nectar disk present.

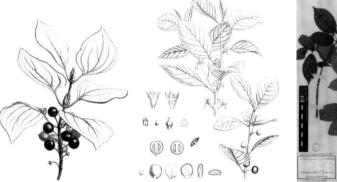

LEFT TO RIGHT:
Rhamnus cathartica;
Frangula alnus;
Frangula crenata.

Characters of similar families: **Cannabaceae**: nectar disk lacking, flowers unisexual. **Celastraceae**: stamens alternating with petals, persistent hypanthium rim lacking at base of the fruit. **Elaeagnaceae**: plant covered in scales, petals always lacking. **Rosaceae**: stamens usually many, petals not enclosing stamens. **Ulmaceae**: trees, leaf veins running into teeth, nectar disk lacking, fruit a samara.

Shrubs, less often **trees** or **climbers**, often armed with thorns or spines, hairs usually simple (stellate in tribe Pomaderreae). **Stipules** usually present and caducous, sometimes spiny (e.g. *Ziziphus* and *Paliurus*). **Leaves** deciduous or evergreen, simple, alternate, sometimes sub-opposite or opposite (e.g. *Rhamnus*, in part), pinnately or 3–5-veined from the base, margins most often serrated, tertiary venation often scalariform. **Inflorescences** cymose, axillary and/or terminal, often congested into fascicles, sometimes flowers solitary or cymes arranged in thyrses and panicles. **Flowers** actinomorphic, usually bisexual, small, <6 mm in diameter, with a hypanthium; calyx lobes 4–5, valvate, keeled adaxially; petals 4–5, usually smaller than sepals, sometimes lacking, usually clawed and hooded; stamens 4–5, free, opposite petals, often enclosed by petals, anthers 2-celled, longitudinally dehiscent, introrse; disk intrastaminal, usually fleshy; ovary superior, half-inferior or inferior, more or less immersed in the disk, (1)2–4 locular; style usually 2–3-cleft; ovule usually 1 per locule. **Fruit** various: indehiscent or splitting, often drupaceous, with 1–several endocarps, sometimes samaroid (e.g. *Paliurus*), persistent hypanthium rim often visible at the base of the fruit. **Seeds** usually 1 per locule.

Literature: Bredenkamp (2000); Chen & Schirarend (2007); Medan & Schirarend (2004); Nesom *et al.* (2016); Tutin (1968).

A family of about 60 genera and 1,200 species, with a near-worldwide distribution but most diverse in Mediterranean-type ecosystems. Members of the family have a range of uses, as edible fruits (*Ziziphus*) and fleshy peduncles (*Hovenia*), dyes (*Rhamnus*), charcoal (*Frangula*), and ornamental shrubs (e.g. *Ceanothus* and *Colletia*).

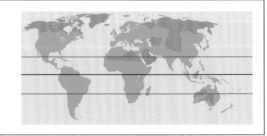

HABIT AND VEGETATIVE PARTS

Shrubs, often armed: **1** *Rhamnus cathartica*, **2** *Rhamnus oleoides*, **3** *Colletia hystrix* and **4** *Ziziphus lotus*. Rarely climbers: **5** *Berchemia scandens*. Hairs stellate in tribe Pomaderreae: **6** *Pomaderris aspera*.

INFLORESCENCES

Axillary fascicles: **7** *Ziziphus lotus*. Sometimes flowers unisexual and petals lacking: **8** *Rhamnus lycioides* (male flowers). Thyrses: **9** *Ceanothus integerrimus*. Head-like clusters: **10** *Phylica dioica*.

FRUITS

Drupes: **11** *Rhamnus alaternus* and **12** *Ziziphus lotus*. Samaroid: **13** *Paliurus spina-christi*. Capsules: **14** *Ceanothus integerrimus*.

95

Cannabaceae

Alison Moore & C. M. Wilmot-Dear

Stipules at petiole base
Leaves simple, alternate
Flowers unisexual
Perianth parts free
Ovary superior

Leaves simple, often 3-veined and asymmetrical at base. **Stipules** caducous. **Flowers** unisexual, in lax axillary unisexual or bisexual cymes or clusters; ovary superior; 2 stigmas; fruit a drupe with persistent stigmas.

LEFT TO RIGHT: *Aphananthe monoica*; *Trema orientale*; *Trema orientale*; *Celtis julianae*.

Characters of similar families: Ulmaceae: leaves not 3-veined from base, flowers bisexual, fruit mostly winged. **Urticaceae**: often herbs, inflorescences congested, stamens inflexed in bud, stigma 1. **Moraceae**: white sap, conspicuous stipules, flowers often in dense heads. **Tiliaceae**: bisexual flowers, petals and sepals, many stamens, stigma 1. **Rhamnaceae**: bisexual flowers with conspicuous disk. **Euphorbiaceae**: stigma usually >2, leaves not 3-veined, veins looped. **Malvaceae–Grewioideae**: bisexual flowers, petals and sepals, many stamens, stigma 1.

Trees, **shrubs**, rarely **herbs**, often spiny. **Stipules**, lateral, free or fused, usually membranous, soon-caducous, rarely persistent. **Leaves** petiolate, alternate, distichous, rarely opposite, simple, rarely palmate-compound (*Cannabis*) or lobed (*Humulus*), often oblique at base; margins entire to toothed; venation conspicuous, pinnate or 3–5-veined from the base; with or without cystoliths. **Inflorescences** axillary, unisexual or bisexual cymes, racemes, clusters or female flowers solitary; plants monoecious, dioecious or polygamous. **Flowers** unisexual, rarely bisexual (e.g. *Celtis*), actinomorphic, 4–5-merous, tepals 1 whorl, usually free and imbricate; male flowers, large, pendulous, conspicuous, wind-pollinated; stamens usually equal and opposite tepals, erect in bud, filaments free; pistillode conspicuous; female flowers sessile, stigmas 2, diverging sometimes bilobed; ovary superior, 1(–2)-locular, ovule 1; with or without staminodes. **Fruit** a drupe, endocarp thick-walled, rarely exocarp winged (only in *Pteroceltis*) or occasionally an achene visible through the persistent, thin, marbled perianth (*Cannabis* and *Humulus*); often with persistent stigma. **Seed** 1.

Literature: Kubitzki (1993b); Manchester (1989); Sherman-Broyles *et al.* (1997); Sytsma *et al.* (2002); Wilmot-Dear (2015a).

A family of 9 genera and ca. 110 species. Widespread throughout tropical, subtropical and temperate regions. The large genus *Celtis* has 65 species in tropical and temperate zones. *Cannabis* and *Humulus* are widely cultivated and naturalised. Ulmaceae subfamily Celtidoideae is now included in Cannabaceae.

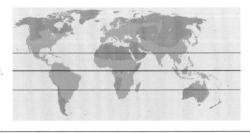

LEAVES

Opposite, simple, lobed: **1** *Humulus lupulus*. Opposite, palmate-compound: **2** *Cannabis sativa*. Alternate, bases oblique and 3-veined: **3** *Celtis koraiensis* and **4** *Celtis jessoensis*.

INFLORESCENCES AND FLOWERS

Axillary clusters: **5** *Trema orientale*. Female inflorescence with bracts: **6** *Celtis glabrata*. Female 'cone' flower: **7** *Humulus lupulus*. Male flower: **8** *H. lupulus*. Male flower with pistillode: **9** *Trema orientale*.

FRUITS

Immature: **10** *Trema orientale*. Drupe with persistent stigmas: **11** *Celtis occidentalis* and **12**, **13** *Celtis australis*. Winged exocarp: **14** *Pteroceltis tatarinowii*. Seeds with reticulate endocarp: **15** *Celtis caucasica*.

Moraceae

Alison Moore & C. M. Wilmot-Dear

Sap or exudate present
Stipules at petiole base
Leaves alternate
Flowers unisexual
Ovary superior

Shrubs, **trees**, **lianas** or **hemi-epiphytes**; white sap; stipules large often leaving circular scar. **Leaves** simple. **Inflorescences** diverse. **Flowers** unisexual, tepals usually 4, often grouped into dense complex heads; perianth 1 whorl; ovary superior; 1–2 stigmas.

LEFT TO RIGHT: *Broussonetia papyrifera; Paratrophis microphylla; Morus rubra; Ficus subpisocarpa.*

Characters of similar families: Urticaceae: no sap; often herbs, cystoliths often elongate; stigma 1, stamens inflexed in bud. **Cannabaceae:** no sap, flowers not in heads, lax inflorescences. **Magnoliaceae:** no sap, flowers large, solitary, bisexual, many stamens and tepals. **Euphorbiaceae:** petiole geniculate, stigmas usually >2, fruit breaks into 3 leaving columella. **Salicaceae** (*Flacourtia*): inconspicuous stipules, many stamens, female flowers with disk.

Shrubs, **trees**, **lianas** or **hemi-epiphytes** (only *Fatoua villosa* is **herbaceous** in the temperate region), sometimes spiny; milky white latex (drying brown, never black). **Stipules** persistent or caducous, often large and amplexicaul leaving a circular scar. **Leaves** alternate (rarely opposite or whorled), simple, entire or toothed (sometimes lobed); venation pinnate (rarely palmate), often conspicuous, 3(–5)-veined from the base; punctiform cystoliths sometimes present. **Inflorescences** pendulous or erect spikes, cymes, racemes, glomerules, or condensed by fusion of flower parts and/or bracts and often enclosed by an involucre; perianth often embedded in and fused with receptacle. Plants monoecious or dioecious. **Flowers** small, unisexual, rarely solitary; tepals (2–)4(–6), free or fused, imbricate or valvate; male flowers with stamens equal and opposite tepals, straight or inflexed in bud, pistillode often present; female flowers with tepals often fused to ovary, sometimes absent; stigmas 1–2 usually filiform; 1(–2) locular, 1 ovule per locule; inferior to superior. **Fruit** often dehiscent drupes, rarely achenes; often grouped into complex disk-like or globose structures or condensed into a syncarp by fusion of parts; specialised infructescence is a syconium; hollow, urceolate receptacle with apical ostiole closed by bracts (fig). **Seeds** 1.

Literature: Brummitt & Wilmot-Dear (2007); Clement & Weiblen (2009); Rohwer & Berg (1993); Wilmot-Dear (2015b).

Family of 45 genera and ca. 1,200 species, cosmopolitan, mostly tropical and subtropical. Evergreen lowland forest, seasonal montane forest (often as pioneer species) or understorey in upland rainforest, hemi-epiphytes in the canopy. The large genus *Ficus* (870 species) contains many economically important fruit trees, which like *Morus* species, are widely cultivated.

LEAVES
Toothed: **1** *Morus alba*. Lobed leaves: **2** *Ficus carica*. Entire with stipule: **3** *F. pumila*. Stipule and stipule scar: **4** *F. carica*.

INFLORESCENCES AND FLOWERS
Male: **5** *Morus alba* and **6** *Paratrophis banksii* (with pistillode). Female (long stigmas):**7** *Maclura pomifera*, **8** *Morus alba*, **9** *Paratrophis smithii* and **10** *P. microphylla*.

FRUITS
Syncarpous aggregate of drupelets: **11** *Morus alba*, **12** *M. alba* with achenes on fleshy stalks and **13** *Broussonetia papyrifera*. Globose heads, seeds embedded in receptacle: **14** *Maclura pomifera*.

FIGS
Syconium: **15** *Ficus rubiginosa* and **16** *F. pumila*. Internal flowers, external bracts: **17** *F. pumila* and **18** *F. carica*.

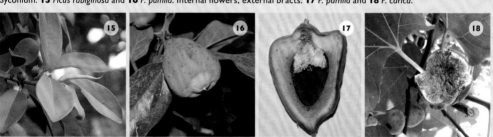

Urticaceae

Alison Moore & C. M. Wilmot-Dear

Stipules at petiole base
Leaves simple
Leaves alternate or opposite
Flowers unisexual
Ovary superior

Leaves simple, often asymmetric, 3-veined, discolorous; cystoliths present; sometimes with stinging hairs. **Flowers** unisexual, minute, in aggregated inflorescences; stamens (1–)3–5, inflexed in bud; stigma 1. **Perianth** 1 whorl. **Fruits** in clusters.

LEFT TO RIGHT:
Urtica pilulifera;
Girardinia diversifolia;
Boehmeria japonica.

Characters of similar families: Cannabaceae: stamens straight, inflorescence always lax, stigmas 2. **Ulmaceae:** stamens straight, inflorescence always lax, stigmas 2, never herbaceous. **Moraceae:** milky sap, stigmas 1–2, cystoliths punctiform or absent, stamens often straight, placentation apical. **Euphorbiaceae:** milky sap, stigmas 3, fruit breaks into 3 leaving central columella. **Melastomataceae:** leaves always opposite, ovary inferior, stipules absent. **Piperaceae:** no cystoliths, perianth absent, lamina entire.

Deciduous herbs, shrubs, subshrubs, (trees or lianas) without latex. **Stipules** often conspicuous and fused, rarely absent. **Leaves** simple; opposite or alternate and spirally arranged; sometimes anisophyllous; leaf form influenced by environment; often asymmetric, 3-veined from base and/or discolorous; margin entire or toothed, sometimes lobed; cystoliths (usually) present, punctiform or elongate; hairs often present, stinging or not. **Inflorescences** unisexual or bisexual (where bisexual unisexual flowers also present), plants monoecious or dioecious; axillary clusters, spikes, panicles, cymes, flattened receptacles or globular fleshy heads, rarely reduced to a single flower; sometimes with involucral bracts; often colourful. **Flowers** unisexual, tiny, perianth (1–)3–5 free or fused tepals; male flowers usually pedicillate; usually actinomorphic, stamens opposite tepals, filaments usually tightly inflexed in bud; explosive pollen release, pistillode 1; female flowers usually sessile; tepals often unequal or completely fused, rarely absent; stigma 1, filiform, capitate or penicillate-capitate; ovary superior; staminodes present or absent; locule 1; ovule 1; pistil 1; placentation basal. **Fruits** often tiny in conspicuous clusters of laterally flattened and asymmetrical achenes; perianth often persistent.

Literature: Friis (1993); Weddell (1869); Wilmot-Dear (2015c); Wilmot-Dear & Friis (2013); Wu *et al.* (2013a).

Family of ca. 55–60 genera and ca. 2,500 species, ca. 20 genera in temperate regions. Worldwide, most numerous in tropics. Mostly found in humid habitats, on forest floors or in riverine vegetation but some genera are adapted to arid environments. Large genera include *Boehmeria*, *Pilea*, *Laportea* (Old and New Worlds), *Elatostema* (Old World), and *Urtica* (temperate regions).

LEAVES

Stinging hairs, stipules: **1** *Urtica ferox*. Cystoliths: **2** *Urtica incisa*. Alternate leaves, asymmetric base: **3** *Elatostema rugosum*. Opposite leaves, 3-veined from base: **4** *Urtica urens*.

INFLORESCENCES

Axillary clusters, male: **5** *Urtica urens*. Axillary clusters, with involucre: **6** *Forsskaolea angustifolia*. Spicate, female: **7** *Boehmeria japonica* var. *tenera*. Cymose panicle, female: **8** *Pilea peperomioides*. Globose heads: **9** *Urtica pilulifera*. Male flowers: **10** *Parietaria judaica*. Stamens inflexed in bud: **11** *Urtica australis*.

FRUITS

Cluster of achenes: **12,13** *Urtica dioica*. Achenes: **14** *Forsskaolea hereroensis* and **15** *U. dioica*.

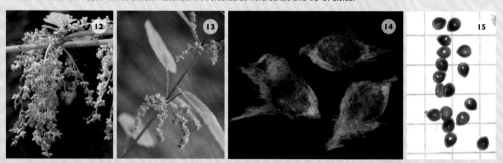

Fagaceae

Alison Moore & C. M. Wilmot-Dear

Stipules at petiole base
Leaves simple
Leaves alternate
Flowers unisexual
Ovary inferior

Trees (or shrubs); monoecious. **Leaves** simple with hairs and/or scales; stipules caducous.
Inflorescences spikes, clusters or solitary. **Flowers** tiny, unisexual, involucrate; ovary inferior; styles 3–6.
Fruits surrounded by a woody cupule or involucre.

LEFT TO RIGHT: *Fagus orientalis; Castanea crenata; Notholithocarpus densiflorus; Quercus marilandica.*

Characters of similar families: Nothofagaceae: peltate stipules, doubly serrate leaf margins. **Betulaceae:** fruits not subtended by cupule. **Juglandaceae:** compound, sometimes opposite leaves, stipules absent. **Myricaceae:** ovary superior, fruit a drupe or nut-like, peltate scales. **Salicaceae:** ovary superior, fruit a capsule. **Lauraceae:** stipules absent, ovary usually superior, cupule absent or fleshy, leaves always entire.

Trees, rarely shrubs. **Stipules** free, caducous. **Leaves** simple, alternate, spirally arranged, rarely opposite or whorled; often scarious, coriaceous; margins entire, toothed or pinnate-lobed; venation pinnate; often with scales and/or abaxially dense hairs, simple or branched, sometimes glandular. **Inflorescences** pendulous (rarely erect) catkins or reduced spikes, heads or cymes, dichasial clusters of 2–30, or reduced to solitary flowers; unisexual or bisexual (female flowers at base of male axes). **Flowers** small, unisexual, plants monoecious, petals absent, perianth bract-like with 4–7(–9) lobes, bracts often caducous. Male flowers rarely solitary; stamens equal and opposite tepals or up to twice in number, rarely more; filaments free, pistillode present or not. Female flowers solitary or in groups of 2–7 along axis, each flower or group surrounded by basal or cup-like involucre (of fused bracts or free scales); perianth usually 6-lobed; ovary inferior; styles and locules 3–6, 2 ovules per locule; pistil 1; staminodes 6–12 or absent. **Fruit** in groups of 1–3(–15) single-seeded nuts surrounded by spiny or scaly, multibracteate cupule, enlarged and hard, forming a dehiscing valvate structure or a basal cup or completely enclosing fruits.

Literature: Camus (1936); Govaerts *et al.* (1998); Kubitzki (1993c); Manos *et al.* (2008); Nixon (1989).

Fagaceae has 8 genera and ca. 1,000 species, mainly in the northern hemisphere; dominant in broad-leaved forests. *Quercus*, as well as *Fagus* and *Castanea*, is a major source of hardwood timber. *Nothofagus* now forms Nothofagaceae, sister to the other Fagales. *Notholithocarpus densiflorus*, the Tanbark Oak (W USA), is now split from *Lithocarpus*.

LEAVES

Entire: **1** *Castanopsis sieboldii*. Lobed: **2** *Quercus cerris*. Toothed: **3** *Quercus agrifolia*. Indumentum stellate: **4** *Quercus ilex*. Indumentum tomentose: **5** *Quercus ilex*. Young leaves and stipules: **6** *Quercus shumardii*.

INFLORESCENCES AND FLOWERS

Male catkins: **7** *Castanea mollissima*, and **8** *Notholithocarpus densiflorus* (with female cupules). Erect catkins: **9** *Lithocarpus henryi*. Pendant, globose head (male): **10** *Fagus sylvatica*. Female flower: **11** *Castanea crenata*.

CAPSULES

Awl-shaped appendages: **12** *Fagus sylvatica*. Saucer-shaped: **13** *Quercus rubra*. Recurved scales: **14** *Quercus acutissima*. Triangular bracts: **15** *Quercus cocciferoides*. Densely spiny: **16** *Castanea sativa* and **17** *Chrysolepis chrysophylla*.

Juglandaceae

Saba Rokni

Stipules absent

Leaves compound

Leaves usually alternate

Flowers unisexual

Ovary inferior

Trees. Leaves pinnately compound, usually alternate, often aromatic, basal leaflets often smaller than terminal ones. **Flowers** small, unisexual, in catkins, spikes or panicle-like inflorescences. **Fruit** a large nut or winged nutlet.

LEFT TO RIGHT:
Juglans regia;
Fruits: 1 *Juglans regia*,
2 *Pterocarya fraxinifolia*,
3 *Juglans nigra*, 4 *Juglans cinerea*, 5 *Carya alba*,
6 *Carya cordiformis* and
7 *Carya illinoinensis*;
Carya sinensis.

Characters of similar families: Betulaceae: leaves simple, stipules present (deciduous), male catkins are often precocious (developing in the previous growing season), the unexpanded catkins hang on the twigs in winter. **Fagaceae:** leaves simple, stipules present (deciduous), fruits enclosed by a multi-bracteate cupule. **Rutaceae:** oily gland dots in leaves, fruit and flowers (crushed material with citrus scent), flowers not in catkins, usually bisexual and 3–5-merous. **Sapindaceae:** leaves without peltate scales and not aromatic, free rachis tip in pinnately compound leaves, flowers not in catkins, ovary superior.

Trees, deciduous or evergreen; bark tight or exfoliating; twigs with solid or chambered pith; buds naked or scaly. **Leaves** alternate or rarely opposite, compound, odd- or even-pinnate, sometimes trifoliolate; stipules absent; leaflets entire or serrate, dotted below with peltate scales, often aromatic, basal leaflets often smaller than terminal ones. **Inflorescences** pendulous or erect, lateral or terminal, several types: androgynous panicles; solitary male catkins; clusters of 3–8 male catkins; 2–several-flowered female spikes. **Flowers** small, unisexual, apetalous, with an entire or lobed bract, sepals 0–4. Male with stamens 3 to >100, in 1–2 or more series, filaments short, anthers basifixed, longitudinally dehiscent. Female with gynoecium of 2–(4) carpels united into an inferior ovary, 1-locular above, but incompletely 2–4(–8)-loculed at base; style 1, short or elongate, rarely absent; stigmas 2–4, plumose or fleshy; ovule 1, orthotropous. **Infructescence** elongate and pendulous, or short and erect, rarely cone-like with persistent bracts. **Fruit** a large nut, with a dehiscent or indehiscent husk, or a 2-, 3- or circular-winged nutlet. **Seed** solitary, without endosperm, cotyledons often fleshy and much folded, germination hypogeal or epigeal.

Literature: Lu *et al.* (1999); Manos & Stone (2001); Plants of the World Online (2021); Stone (1993); Stevens (2001 onwards).

Nine genera and 71 species; mostly in temperate and subtropical regions of the northern hemisphere, also Malesia, Central America and the Andes. Temperate genera are *Juglans*, *Pterocarya*, *Carya*, *Platycarya*, and *Cyclocarya*. Some genera produce edible nuts (walnut, pecan, and hickory) and timber. *Rhoiptelea chiliantha* (formerly Rhoipteleaceae) differs in some key characters and is excluded from the description.

HABIT, TWIGS AND LEAVES

Trees: **1** *Juglans nigra*. Twig with chambered pith: **2** *Juglans regia*. Leaves pinnately compound, basal leaflets often smaller than terminal: **3** *Juglans regia* (leaflets entire) and **4** *Platycarya strobilacea* (leaflets serrate).

INFLORESCENCES

Solitary male catkins: **5** *Pterocarya stenoptera* and **6** *Juglans regia*. Female spikes: **7** *Pterocarya fraxinifolia* (pendulous) and **8** *J. regia* (erect). Erect androgynous panicle, male catkins flanking the male and female catkin in the centre: **9** *Platycarya strobilacea*.

INFRUCTESCENCES AND FRUITS

Infructescence cone-like with persistent bracts: **10** *Platycarya strobilacea*. Infructescence elongate and pendulous, fruit a winged nutlet: **11** *Pterocarya fraxinifolia* (2-winged) and **12** *Cyclocarya paliurus* (circular-winged). Infructescence short and erect, fruit a large nut enclosed in a husk: **13** *Juglans nigra* (indehiscent husk) and **14** *Carya ovata* (dehiscent husk).

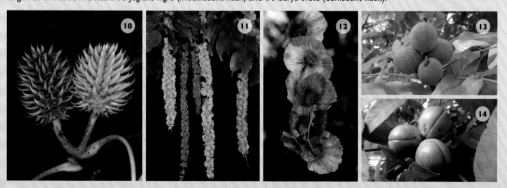

Betulaceae

Saba Rokni

Stipules present
Leaves simple
Leaves alternate
Flowers unisexual
Ovary inferior

Deciduous **trees** and **shrubs**. **Leaves** simple, alternate, usually serrate. Monoecious. **Flowers** much reduced, in unisexual inflorescences. Male catkins pendulous, cylindric, conspicuously bracteate, often precocious. **Fruit** a nut, nutlet or samara.

LEFT TO RIGHT:
Betula pubescens: 1,2 male flowers, 3 female catkin, 4 female flowers, 5 fruit (tiny samara), 6 fruit and scale, and 7 seed;
Alnus incana subsp. *tenuifolia*: 1 male flowers, 2 anthers, 3 infructescence section, 4 fruit and scale;
Betula grossa.

Characters of similar families: Juglandaceae: leaves compound and often aromatic, stipules absent. Fagaceae: fruits enclosed by a multi-bracteate cupule. Salicaceae: plants usually dioecious, teeth of leaves often with a spherical or papillate gland at the apex, ovary superior, fruit a berry or capsule.

Trees and **shrubs**, deciduous, monoecious; bark close or exfoliating in thin layers, often with prominent lenticels. **Winter buds** stalked or sessile, scaly or naked. **Stipules** free, deciduous. **Leaves** alternate, simple, petiolate, pinnately veined, sometimes lobulate, margin serrate to nearly entire. **Inflorescences** terminal or lateral, unisexual, flowers much reduced. **Male catkins** often precocious, pendulous, cylindric, conspicuously bracteate, consisting of crowded, 1–3-flowered clusters; stamens (1–)4–6; filaments very short; anthers 2-locular, medifixed, dehiscing by longitudinal slits, thecae connate or separate. **Female inflorescences** either of erect to pendulous bracteate catkins, or of compact clusters of tiny, highly reduced flowers; gynoecium 2(–3)-carpellate; ovary inferior, usually 2-locular below, 1-locular above; placentation axile; ovules 1–2 per locule, pendulous; styles 2, linear, free. **Infructescences** cone-like with small, crowded, woody or leathery scales, these are deciduous with the fruits or persistent; or formed of clusters subtended or enclosed by large, nearly leafy bracts falling with the fruits. **Fruit** a nut, nutlet, or tiny 2-winged samara. **Seed** 1, endosperm present but thin at maturity, embryo straight, cotyledons flat or greatly thickened, oily.

Literature: Kubitzki (1993a); Furlow (1997); Li & Skvortsov (1999); Plants of the World Online (2021); Stevens (2001 onwards).

Six genera (*Alnus*, *Betula*, *Carpinus*, *Corylus*, *Ostrya* and *Ostryopsis*) and 150–200 species; mainly in cool temperate and boreal northern hemisphere, some on tropical mountain ranges through Central America and the Andes to Argentina, or in SE Asia. *Ostryopsis* is restricted to China. Used for timber, charcoal, hazelnuts (*Corylus*), ornamental trees and shrubs, and soil nitrification (*Alnus*).

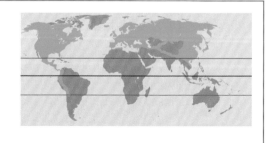

HABIT, BARK, LEAVES AND BUDS

Tree: **1** *Betula pendula*. Shrub: **2** *Corylus avellana*. Bark peeling in thin layers with prominent lenticels: **3** *Betula ermanii*. Winter bud stalked: **4** *Alnus glutinosa*. Leaves simple, alternate and serrate: **5** *Carpinus betulus*.

INFLORESCENCES

Winter catkins (female above male): **6** *Alnus glutinosa*. Spring catkins: **7** *Betula ermanii* (female erect); **8** *Ostrya virginiana* (female left, pendulous, with pink styles); **9** *Alnus cordata* (female). Compact cluster of female flowers (enclosed in bud scales, red styles exposed): **10** *Corylus avellana*.

INFRUCTESCENCES AND FRUITS

Infructescences cone-like: **11** *Betula pendula* (scales falling with fruits); **12** *Alnus cordata* (scales woody and persistent). Infructescences with almost leafy bracts: **13** *Carpinus betulus* (nutlets); **14** *Ostrya carpinifolia* (nutlets completely enclosed by bracts); **15** *Corylus avellana* (nuts).

Cucurbitaceae

Anna Trias-Blasi

Stipules absent
Leaves alternate
Leaves simple or compound
Flowers unisexual
Ovary inferior

Usually **climbers** with tendrils at 90° from the petiole. **Leaves** simple, often lobed or palmately compound. **Leaflets** often roughly hairy. **Inflorescences** usually axillary. **Flowers** unisexual, ovary inferior. **Fruit** a berry, hard or soft-skinned, or a fleshy or dry capsule.

 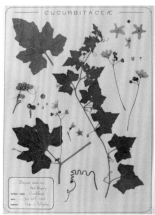

LEFT TO RIGHT:
Cucurbita pepo: note simple leaves, tendril 90° from the petiole, axillary inflorescence and unisexual flowers; *Bryonia dioica*.

Characters of similar families: Vitaceae: leaf-opposed inflorescence, ovary superior. **Convolvulaceae**: tendrils absent, tubular bisexual flower, ovary superior. **Smilacaceae**: tendrils in pairs, often prickles present on stems and/or leaves, inflorescence an umbel, flowers inconspicuous, ovary superior.

Usually **climbers**, sometimes scrambling, rarely decumbent shrubs. Climbing by a simple or branched tendril arising 90° to the petiole base. **Stipules** absent. **Leaves** alternate, simple, often lobed, sometimes palmately compound, venation generally palmate, usually with rough hairs and with a foetid smell. **Inflorescence** solitary or in cymes, axillary. **Flowers** unisexual in monoecious and dioecious plants, actinomorphic; calyx (3–)5(–7)-merous with lobes arising from a hypanthium; corolla (3–)5(–7)-merous usually white to yellow and orange, rarely red or pink; stamens 1–5, attached to the top of the hypanthium, anthers dehiscing longitudinally; ovary inferior with 3 fused carpels, usually with parietal placentation or 2–5 locular by intrusive placentas; style usually 1, sometimes 3, generally stout, stigmas bifid. **Fruit** usually a hard-skinned berry, sometimes small soft-skinned small berries, sometimes a fleshy capsule, either dehiscent or indehiscent, 1–many-seeded. **Seeds** usually tear-shaped and flattened, sometimes with irregular margins and varied ornamentation, sometimes winged, embryo oily and non-endospermic.

Literature: De Wilde & Duyfjes (2008); Jeffrey (1980);. Taylor & Zappi (2009).

Cucurbitaceae comprises ca. 800 species in 120 genera. It is mostly tropical and subtropical, but some genera are completely or partially distributed in temperate areas. Various species are of major importance as a source of food and they are widely cultivated for their fruits, which include cucumbers, courgettes, melons and pumpkins.

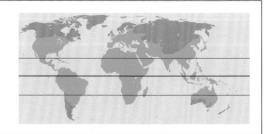

HABIT, TENDRIL AND LEAVES

Climbing habit: **1** *Thladiantha dubia*. Tendril position and male flower: **2** *Momordica charantia*. Leaf shape cordate: **3** *T. dubia*.

INFLORESCENCES

Male flower: **4** *Ecballium elaterium*. Female flower: **5** *Bryonia dioica*. Inferior ovary: **6** *Cucumis sativus*.

FRUIT

Dehiscent capsule, flattened seeds: **7** *Momordica charantia*. Dehiscent capsule, papery seeds: **8** *Alsomitra macrocarpa*. Berry: **9** *Cayaponia tubulosa*. Indehiscent fleshy capsule: **10** *Cucumis sativus* and **11** *Benincasa hispida*.

Celastraceae

Timothy Utteridge

Leaves simple
Leaves alternate or subopposite
Flowers actinomorphic
Perianth parts free
Ovary superior

Leaves simple, usually alternate or subopposite, often toothed and drying pale grey-green. **Flowers** usually small and white to greenish, disk present, stamens opposite the calyx-lobes, petals free. **Fruit** various, capsular to indehiscent, often with a coloured aril.

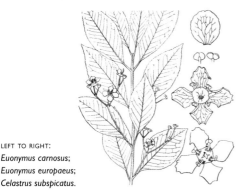

LEFT TO RIGHT:
Euonymus carnosus;
Euonymus europaeus;
Celastrus subspicatus.

Characters of similar families: **Aquifoliaceae:** leaves often drying blackish, disk absent, stigma broad and sessile, fruit with several pyrenes. **Rhamnaceae:** leaves not subopposite, tertiary venation scalariform, stamens opposite the petals. **Rubiaceae:** stipules interpetiolar, leaves opposite with entire margins, corolla tubular, ovary usually inferior. **Salicaceae:** leaves alternate often with distinctive 'salicoid teeth', stamens usually many, ovary 1-locular.

Trees, **shrubs**, **herbs** or **lianas**. **Stipules** small, caducous, or absent. **Leaves** simple, often subopposite, alternate, spiral, decussate or opposite, penninerved, margins serrate, crenate or entire. **Inflorescences** axillary and/or terminal, extra-axillary, variously branched. **Flowers** small, actinomorphic, bisexual or unisexual (plants usually dioecious), white to greenish; calyx 4- or 5-lobed, lobes imbricate, rarely valvate, usually persistent; petals 4 or 5, free, rarely absent, imbricate, contorted, rarely valvate, caducous, sometimes persistent; stamens (2–)3, 4, or 5, alternate with the petals, filaments inserted on or within the disk; anthers 2-celled, usually longitudinally dehiscent; disk various, often present and conspicuous, fleshy or membranous, entire to lobed, usually smooth or with papilla-like processes; ovary superior or partly or entirely immersed in the disk, (1–)2–5(–many)-celled; styles 0, 1 or 3, distinct, short, or lacking, terminal; stigma(s) simple, or lobed; ovules usually 1–2 in each cell, sometimes numerous, anatropous. **Fruit** various: capsular, loculicidal or with 3 divergent separate or laterally connate 'follicles', or drupaceous, and indehiscent. **Seeds** erect or pendulous, sometimes winged; aril present or none, usually orange or orange-red, rarely white, when present.

Literature: Jessup (1984); Ma *et al.* (2008, 2016); Simmons (2004).

A family of ca. 100 genera with ca. 1,300 species. Cosmopolitan genera with the largest numbers of species in temperate regions include *Euonymus* (140 species) and *Maytenus* (ca. 175 species). The herb genera *Parnassia* (ca. 55 species) and *Lepuropetalon* (1 species) are now considered to be members of this family.

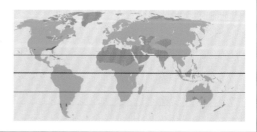

LEAVES

Leaves entire: **1** *Maytenus oleoides*. Subopposite, simple leaves with finely serrate margins: **2,3** *Euonymus hamiltonianus*.

FLOWERS

Terminal inflorescences: **4** *Celastrus orbiculatus*. Flowers with conspicuous disk and stamens alternating with petals: **5** *Gymnosporia royleana* and **6** *Euonymus japonicus*. White petals with distinct veins and 4 stigmas: **7** *Parnassia palustris*.

FRUIT

Mature fruits before dehiscence: **8** *Celastrus orbiculatus* and **9** *Denhamia bilocularis*. Seeds with arils: **10** *Euonymus europaeus* and **11** *C. orbiculatus*.

Oxalidaceae

Martin Xanthos

Leaves alternate
Leaves pinnately or palmately compound
Flowers actinomorphic
Trimorphic heterostyly in many species
Fruit a capsule

Leaves often pinnately or palmately compound, leaflets articulate. **Flowers** actinomorphic, often bright and showy; ovary superior. **Fruit** a capsule, seeds of some species may have a fleshy aril at the base.

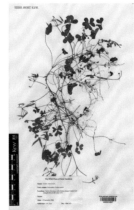

LEFT TO RIGHT:
Oxalis articulata: note 2 whorls of stamens; *Oxalis acetosella*; *Oxalis corniculata*.

Characters of similar families: Connaraceae: fruit follicular, usually splitting down one side. Cunoniaceae: leaves usually opposite, flowers not usually bright and showy. Elaeocarpaceae: perianth segments valvate, stamens not arranged in two whorls, filaments free.

Perennial herbs (rarely annual), **small trees**, **shrubs** or **climbers**, usually clump forming or spreading by stolons, often with underground storage bulbs, tubers or fleshy roots. **Leaves** alternate but often pinnately or palmately compound (often trifoliate in *Oxalis*), from a well-defined petiole, leaflets articulate, occasionally leaves replaced by phyllodes (e.g. *Oxalis fruticosa*); petioles sometimes woody and persistent, rarely succulent. **Inflorescences** thyrsopaniculate, forming an umbel, spike or head. **Flowers** often bright and showy, rarely apetalous and cleistogamous, actinomorphic, hermaphrodite; sepals 5, persistent and overlapping, petals 5, usually twisted in bud, free or fused at the base, clawed, often brightly coloured; stamens 10, arranged in 2 whorls, connate at the base, outer whorl of 5 lying opposite the petals, styles 5, free, styles can be of different lengths in flowers on the same plant (heterostyly), stigmas capitate or 2-cleft, ovary superior, carpels 5, free or fused, 5-locular, each locule with 1 or 2 rows of ovules, placentation axile. **Fruit** a capsule; seeds sometimes with a fleshy aril at the base (some *Oxalis* and *Biophytum* species), which results in explosive dehiscence.

Literature: Christenhusz *et al.* (2017); Heywood *et al.* (2007); World Checklist of Selected Plant Families (2021).

An almost cosmopolitan family of 5 genera and ca. 570 species. Most of the family occurs in high altitude zones in the tropics; the remainder is found in subtropical or sometimes temperate areas. The largest genus is *Oxalis* (ca. 500 species). The leaves of some *Oxalis* species are used in salads.

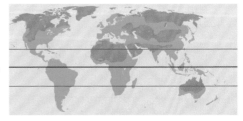

HABIT AND LEAF ARRANGEMENT

Usually herbs with trifoliate leaves: **1** *Oxalis triangularis*, **2** *Oxalis cytisoides* and **3** *Oxalis confertifolia*.

FLOWER MORPHOLOGY

Actinomorphic, showy flowers: **4** *Oxalis articulata*, **5** *Oxalis melanosticta*, **6** *Oxalis massoniana*, **7** *Oxalis squamata*, **8** *Oxalis obtusa*, **9** *Oxalis semiloba* and **10** *Oxalis corniculata*.

FRUITS

Capsules: **11** *Oxalis dillenii* and **12** *Oxalis acetosella*.

Violaceae

Sue Zmartzy & Harvey Ballard

Leaves simple
Leaves alternate
Flowers bisexual
Perianth parts free
Superior ovary

Trees, shrubs, lianas, herbs. **Leaves** alternate, simple, commonly stipulate. **Flowers** usually bisexual, slightly to strongly zygomorphic, 5-merous, anther connective well-developed, often with scale, ovary superior, 1-locular, placentation parietal, style 1. **Fruit** commonly a capsule or berry.

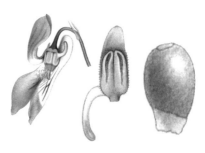 

LEFT TO RIGHT:

Viola riviniana: note the spur, stamen cone, anther scale (orange), nectary gland within spur, and seed with elaiosome;

Viola disjuncta: note the large, deeply lobed stipules.

Characters of similar families: Lentibulariaceae (*Pinguicula* with *Viola*): leaves sticky, anthers 2. **Ericaceae** (*Leucopogon* with *Melicytus*): exstipulate, sympetalous, stamens inserted at top of floral tube, fruit a drupe. **Primulaceae-Myrsinoideae** (*Myrsine*, *Rapanea*, with *Melicytus*): exstipulate, stamens antepetalous, staminodes 0, fruit a drupe. **Euphorbiaceae** (with 3-locular 3-valved capsules): capsule has central columella persistent in fruit, placentation axile or basal, 1 seed per carpel.

Trees, shrubs, **lianas**, caulescent or acaulescent **herbs**. **Hairs** often present, simple. **Stipules** present. **Leaves** usually alternate (rarely opposite), simple, entire, lobed or dissected, mostly lanceolate to ovate; margins glandular-toothed. **Inflorescence** axillary or terminal; fasciculate, cymose, racemose, thyrsoid or paniculate, or flowers solitary; pedicels articulated (except *Viola*). **Flowers** commonly bisexual, slightly to strongly zygomorphic (mostly expressed in corolla); sepals 5, free, subequal, occasionally prominently unequal, in *Viola* each with a basal appendage (auricle); petals 5, free, equal or unequal, anterior petal often differentiated, dilated and/or spurred; stamens 5, antesepalous, free or connate into a tube closely encircling ovary ('pollen cone'), anther connective often well-developed and terminating in a scale; in more zygomorphic flowers 2 filaments and/or anthers bear nectariferous glands extending into petal spur; ovary superior, 1-locular, carpels commonly 3, placentation parietal, ovules 1–many; style terminal, simple (often lobed or hooked). **Fruit** a 3-valved capsule, less often a dry or fleshy berry, nut, follicle or papery bladder. **Seeds** globose to ovoid, less often flattened, winged or angled, commonly bearing an elaiosome.

Literature: Ballard *et al.* (2014); Breitwieser *et al.* (2010–2021); Little & McKinney (2015); Utteridge (2015); Wahlert *et al.* (2014).

25 genera, 1,100 species. Worldwide, mostly pantropical. The three largest genera are primarily herbaceous *Viola*, mostly woody *Rinorea*, and herbaceous and suffruticose *Hybanthus* (polyphyletic, being dismantled). At low to high altitude in wet or dry habitats. Three predominantly temperate genera: *Viola* (525 species), *Melicytus* (10 species) and *Cubelium* (1 species). *Viola* includes many ornamentals.

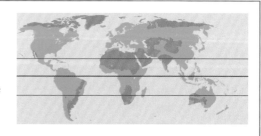

HABIT, LEAVES, INFLORESCENCES

Herbs (rosette habit): **1** *Viola pluviae*. Leaves lobed: **2** *Viola pedata*. Columnar, succulent: **3** *Viola turritella*. Herbs (alternate leaves, axillary cymes): **4,5** *Cubelium concolor*. Shrub: **6** *Melicytus ramiflorus*, and **7** *Melicytus lanceolatus* (with axillary flowers).

FLOWERS

Strongly zygomorphic: **8** *Viola epipsila* (note stamen cone with orange anther scales). **9** *Viola cornuta* (note petal spur). Weakly zygomorphic: *Melicytus ramiflorus* (**10** ♀, **11** ♂) (note nectar droplets from stamens or staminodes).

FRUIT

Capsule, pre-dehiscence: **12** *Viola macloskeyi*. Capsule, dehisced, with seed attached to capsule walls: **13** *V. macloskeyi* and **14** *Cubelium concolor*. Berries: **15** *Melicytus ramiflorus*.

Salicaceae

Sue Zmarzty

Pellucid glands or other foliage glands present
Leaves simple
Leaves alternate
Flowers actinomorphic
Ovary superior

Trees, **shrubs**. **Leaves** simple, alternate, often with glands, often toothed. **Flowers** often small, petals free, generally isomerous with sepals or absent; disk or disk glands typically present; ovary superior, 1-locular, placentation parietal, rarely basal. **Fruit** a berry or capsule, seed often arillate.

LEFT TO RIGHT:
Top: *Salix* ♂ and ♀ flowers, and seed; middle: *Populus* ♂ and ♀ flowers, and bract; bottom: spherical leaf tooth gland and petiole apex glands. *Salix rostrata* var. *luxurians*.

Characters of similar families: Betulaceae (including Corylaceae): leaf margin doubly serrate, staminate flowers often in cymules, fruit a nut. **Celastraceae**: stamens 3–5, alternate with petals, ovary rarely 1-locular. **Euphorbiaceae** and **Phyllanthaceae**: flowers unisexual, ovary usually 3-locular, placentation axile-apical. **Rhamnaceae**: calyx tubular at base, petals usually present, usually hooded enclosing anther, stamens 4–5, ovary rarely 1-locular. **Rhizophoraceae** (only *Cassipourea flanaganii* or *C. malosana*, confused with *Pseudoscolopia*): stipules interpetiolar, petal apex divided.

Trees or **shrubs**. **Hairs** usually simple. **Stipules** generally present. **Leaves** simple, alternate, rarely opposite, sometimes pellucid-punctate or -striate, usually toothed, tooth apex often with spherical (e.g. temperate genera) or papillate gland, venation pinnate or palmate. **Inflorescence** axillary or terminal, flowers solitary or in fascicles, cymes, corymbs, racemes, catkins (e.g. *Salix* and *Populus*) or panicles. Flowers bisexual or unisexual, actinomorphic, often small; sepals 0 (*Salix* and *Populus*) or (2–)3–5(–22), then free or partly fused, rarely conduplicate, petals 0 (e.g. most temperate genera), or generally isomerous with and similar to sepals, free; disk glands or nectary scales often present, at stamen bases, inter- or extra-staminal or -gynoecial, or a cup adnate to inside of calyx, or a single disk bearing gynoecium or stamens; stamens (1–)2–150, free, often exserted, sometimes on disk rim alternating with disk-lobes; ovary superior, rarely semi-inferior, 1-locular, rarely falsely divided, placentation parietal, rarely basal (*Salix* or *Populus*). **Fruit** usually a capsule (rarely with outer layer shedding, and placentae woody, persistent), or a berry. **Seed** 1–many, often arillate, rarely winged, or with a coma of hairs (*Salix* and *Populus*).

Literature: Chase *et al.* (2002); Hodel & Henrich (2020); Killick (1976); Sleumer (1977); Yang & Zmarzty (2007).

About 55 genera, 1,200 species. Cosmopolitan, with 8 predominantly temperate genera: *Salix* (ca. 450 species) and *Populus*, which form dominant and extensive northern forest and are economically important; *Azara* in temperate South America; *Pseudoscolopia* in South Africa (Near Threatened); and *Idesia*, *Poliothyrsis*, *Carrierea* and *Itoa* in mixed forest in temperate and northern subtropical regions of Asia.

HABIT, LEAVES, INFLORESCENCES, FLOWERS

Tree, deltoid leaf, ♀ catkin in fruit: **1** *Populus* cf. *fremontii*. ♂ catkin: **2** *Populus deltoides*. Opposite leaves, ♂ flowers, petals and sepals alike: **3** *Pseudoscolopia polyantha*. Reduced flowers, erect catkins: **4** *Salix repens* (♂) and **5** *Salix hastata* (♀). Axillary cyme, flowers, ♂ petals 0: **6** *Azara* cf. *dentata*. Large calyx, sepals conduplicate, petals 0: **7** *Carrierea calycina* (♂). Petals 0, extragynoecial disk: **8** *Idesia polycarpa* (♀). Leaf large, oblong, pinnate-veined: **9** *Itoa orientalis*.

FRUIT

Dehiscent capsules with cottony hairs, seed: **10** *Salix nigra*. Capsule, persistent inner layer and woody placentae, winged seed (not to scale): **11** *Itoa orientalis*. Berry: **12** *Azara microphylla* (axillary) and **13** *Idesia polycarpa* (in panicles).

Euphorbiaceae *s.s.*

Gill Challen & Laura Pearce

Milky latex often present
Stipules present
Leaves simple, alternate
Flowers unisexual
Ovary superior

Latex often produced. **Leaves** usually simple, alternate, stipulate. **Flowers** unisexual, often reduced. **Ovary** superior. **Fruit** usually a 3-locular schizocarp with a central columella and 1 ovule per locule.

LEFT TO RIGHT:
Mercurialis perennis: note opposite leaves, stipules, absence of petals and bilocular fruit.
Euphorbia squamigera: note entire, alternate leaves.

Characters of similar families: Amaranthaceae: stipules lacking, flowers bisexual, fruit an achene, utricle or pyxis. **Cactaceae:** latex and cyathia absent (vs. *Euphorbia*), areoles present, ovary inferior, fruit a many-seeded berry. **Crassulaceae:** stipules lacking, flowers bisexual, fruit an aggregate of follicles with several seeds in each. **Salicaceae:** flowers subtended by a hairy bract (*Salix* and *Populus*), styles not divided, ovary 1-locular, placentation parietal or basal, ovules 1–numerous, seeds often with basal tuft of hairs. **Urticaceae:** leaves with cystoliths, carpels 1 (stigma usually solitary), fruit an achene, small drupe, or multiple fruit.

Herbs, **shrubs**, **trees**, **climbers** or **succulents**, with white (rarely coloured) latex, or clear sap. **Hairs** simple, stellate, lepidote, dendritic, or T-shaped. **Stipules** present, rarely absent, sometimes glandular. **Leaves** alternate, rarely opposite or whorled, simple, margins entire to dentate or palmately lobed, venation pinnate to palmate or reduced, glands often present on leaf blade or petiole. **Inflorescences** unisexual or bisexual (plants monoecious or dioecious), axillary or terminal, rarely cauliflorous, spicate, racemose, paniculate, fasciculate-glomerulate, reduced to a cyathium (a cup-like structure composed of bracts, glands and several naked flowers) or rarely solitary. **Flowers** unisexual, usually small, actinomorphic, rarely zygomorphic; sepals 3–6(–8) free to fused, imbricate to valvate, rarely absent; petals 3–6 or absent; disk present or absent; stamens 1–many, filaments free, fused or absent, anthers bilocular, longitudinally dehiscent; ovary superior, smooth to spiny, syncarpous 2–3(20) locules; styles free or fused, entire or bifid to multifid; ovules 1 per locule. **Fruit** usually a dry dehiscent, 3-locular capsule (schizocarp) with cocci separating from a central, often persistent columella; rarely an indehiscent fleshy drupe. **Seeds** smooth or sculptured; carunculate or not.

Literature: Efloras.org (2021); Govaerts *et al.* (2000); Radcliffe-Smith (2001); Webster (1994); Wurdack *et al.* (2005).

The largest of 7 families to be segregated from a much larger family, Euphorbiaceae *sensu lato*. Euphorbiaceae s.s. has ca. 220 genera and 6,500 species, including the 'giant' genera *Euphorbia* and *Croton*. Distribution cosmopolitan, except in the polar regions. Economically important representatives include *Hevea brasiliensis* (rubber) and *Manihot esculenta* (cassava).

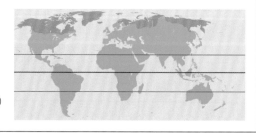

HABIT AND LATEX

Herbs: **1** *Euphorbia helioscopia* and **2** *Euphorbia characias*. Shrubs: **3** *Bertya tasmanica*. Alternate, simple leaves and white latex: **4** *Euphorbia helioscopia*.

INFLORESCENCES

Cyathium: **5** *Euphorbia colorata*. Thyrse, female flowers at base and male flowers above: **6** *Neoshirakia japonica*. Inconspicuous female flowers, petals absent: **7** *Amperea xiphoclada*. Male flowers: **8** *Bertya tasmanica*.

FRUITS

Fruit with multifid styles: **9** *Croton texensis*. 3-locular fruits with 1 seed per locule, seeds attached to apex of columella: **10** *Colliguaja* sp. and **11** *Euphorbia lathyris*. Columella: **12** *E. lathyris*.

Linaceae

Sally Dawson

Stipules present or absent
Leaves alternate or opposite
Leaves simple
Leaf margins entire or serrate
Ovary superior

Herbs, trees, shrubs or **lianas. Leaves** opposite or alternate. **Flowers** petals free or fused at base; style 3–5 free or forming a column; stamens fused at base. **Seeds** flattened.

LEFT TO RIGHT: *Linum grandiflorum*: note style 5, forming a column and free sepals; *Linum usitatissimum*: note superior ovary and filaments fused at base; *Tirpitzia sinensis*; *Anisadenia saxatilis*: note inflorescence a raceme.

Characters of similar families: Ixonanthaceae: leaves alternate, single simple style, disk present, seeds arillate or winged. **Ctenolophonaceae:** single style, bifurcate at apex, stamens free, disk present, 1-seeded, arillate. **Humiriaceae:** simple single style, disk present, sap present.

Herbs, trees, shrubs or **lianas** climbing by hooks or woody tendrils; glandular on floral and vegetative organs. **Leaves** simple alternate, whorled or can be opposite, venation sometimes parallel or reticulate, margins entire or can be serrate, stipules present or absent, sometimes modified into glands. **Inflorescence** racemes, panicles, terminal spikes or axillary fascicles or flowers solitary. **Flowers** bisexual, actinomorphic, hypogynous, sometimes heterostylous, sepals (4)5, free or partially fused at base, sometimes unequal; petals larger than sepals in late bud, protecting the flower organs, (4)5, free sometimes clawed, with a narrow base or fused at base to form a tube, with or without appendages, usually contorted in upper part, valvate or joined below; stamens 5–10(15), unequal, alternating or opposite sepals, with filaments fused at base, often with staminodes, and extra staminal nectary glands; anthers longitudinally dehiscent; ovary superior, usually 2–5-locular, pendulous ovules (1)2 per locule, septa and sometimes a false septum partially to fully dividing locules, placentation axile, style usually 3–5, sometimes tristylous, free or forming a column, stigmas sometimes capitate. **Fruit** a drupe, mericarp or septicidal capsule. **Seeds** flattened, arillode slight or none, sometimes mucilaginous, endosperm abundant, scarce or absent.

Literature: Endress *et al.* (2013); Matthews & Endress (2011); McDill *et al.* (2009); Liu & Zhou (2008).

A cosmopolitan family of 9 genera and ca. 280 species. The position of the genera and the family within the Malpighiales is not fully resolved. The main genera in temperate regions are those in the subfamily Linoideae. *Linum usitatissimum* is a major crop cultivated for oil (linseed oil) and fibre (flax) to make linen.

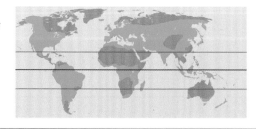

HABIT

Usually herbs: **1** *Linum amurense*. Leaves alternate: **2** *Linum usitatissimum*. Leaves opposite: **3** *Radiola linoides*. Sometimes low-growing or spreading shrubs: **4** *Reinwardtia indica*.

INFLORESCENCES AND FLOWERS

Panicles: **5** *Linum tenue*, **6** *Linum tenuifolium* and **7** *Linum corymbulosum*. Stamens opposité petals, exserted filaments, exserted style arms: **8** *L. tenuifolium*. Stipular glands and clawed petals: **9** *Linum campanulatum*. Glandular sepals: **10** *Anisadenia pubescens*. Contorted corolla: **11** *Linum suffruticosum*.

FRUIT AND SEED

Developing capsules: **12** *Linum amurense*. Style remaining: **13** *Reinwardtia indica*. Dried fruit: **14** *Linum usitatissimum*. Open fruit: **15** *Linum* sp. Seeds flattened: **16** *L. usitatissimum*

Geraniaceae

Nina Davies

Stipules present
Leaves simple or compound
Leaves alternate or opposite
Flowers bisexual
Ovary superior

Annual or perennial **herbs**, stipules present. **Leaves** pinnately or palmately lobed or compound, alternate or opposite. **Flowers** usually bisexual, actinomorphic or zygomorphic. **Fruit** a schizocarp with 1-seeded awned mericarps.

LEFT TO RIGHT:
Pelargonium multibracteatum;
Geranium endressii;
Erodium carvifolium.

Characters of similar families: Francoaceae: occur mainly in South America and Africa; shrubs, small trees, rarely herbs. **Malvaceae:** usually stellate hairs on leaves and stem; stamens usually fused into a tube. **Montiaceae:** stipules absent, leaves in a basal rosette, simple. **Oxalidaceae:** stipules usually absent, corolla lobes contorted, stamens fused into a tube.

Annual or perennial **herbs**, rarely shrublets, shrubs or geophytes with root tubers. **Stipules** present or absent (*Hypseocharis*). **Leaves** alternate or opposite, mostly palmately or pinnately lobed, dissected or compound, rarely spinescent; serrate, crenate or dentate (rarely entire); petiolate. **Inflorescence** axillary, cymose, pseudoumbellate or rarely solitary. **Flowers** usually bisexual, actinomorphic or zygomorphic; sepals 5, usually distinct, imbricate with valvate tips; petals free, usually 5 (4, 2 or 0), imbricate; fertile stamens 5, 10 or 15, usually in 2 whorls, sometimes a whorl reduced to staminodes; filaments basally connate or distinct; anthers 2-thecous, dehiscing longitudinally, introrse; styles present or absent, if present united in flower, stigmas usually 5, filiform, ligulate or clavate; ovary superior; 3–5 (rarely 2 or 8)-locular, connate; ovules 1–2(–12) per locule, pendulous, anatropous; placentation axile; nectaries usually 5 and alternate with petals or united in a tube. **Fruit** a schizocarp with 1-seeded (rarely 2–many-seeded) awned mericarps that separate elastically from a central beak (fruits can be a ventricidal capsule in *Hypseocharis*) or sometimes a 3–5 (rarely 8)-lobed. **Seeds** usually with little or no endosperm, or embryo folded or a cochlear embryo with spirally folded cotyledons (*Hypseocharis*).

Literature: Albers & Van der Walt (2007); Plants of the World Online (2019); Stevens (2001 onwards); Xu Langran & Aedo (2008).

A family of 8 genera and ca. 800 species. Widely distributed in temperate and sub-tropical climates. *Geranium* is the largest genus with ca. 350 species, followed by *Pelargonium* with ca. 280 species. The common name 'cranesbills' is derived from the shape of the schizocarp before the mericarps are released. Pelargoniums, geraniums and erodiums are popular ornamentals.

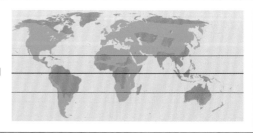

HABIT AND LEAVES

Usually herbs: **1** *Geranium robertianum*, **2** *Geranium nodosum* and **3** *Pelargonium denticulatum*. Leaves palmately divided: **4** *Geranium molle*. Leaves pinnately compound: **5** *Erodium moschatum*.

FLOWERS

Flowers actinomorphic: **6** *Erodium corsicum* (styles present and united in flower, stigmas 5, stamens 5, alternating with 5 staminodes), **7** *Geranium pratense* and **8** *Geranium tuberosum* (stamens 10), and **9** *Monsonia angustifolia* (stamens 15). Flowers zygomorphic, styles present and united in flower, stigmas 5, stamens 10, 3–8 staminodal: **10** *Pelargonium radens*.

FRUIT

Fruit a schizocarp, 1-seeded (rarely 2–many-seeded) awned mericarps attached to a central beak: **11** *Erodium manescavi*, **12** *Monsonia angustifolia* and **13** *Geranium robertianum*. Mericarps separating from the beak: **14** *Geranium collinum* and **15** *Geranium rotundifolium*.

123

Lythraceae

Eve Lucas

Leaves simple
Leaves alternate or opposite
Flowers bisexual
Perianth parts fused
Ovary superior or semi-inferior

Herbs or **sub-shrubs**. **Leaves** simple, opposite, alternate or whorled, margins entire. **Calyx** tubular with free lobes, petals often clawed, inserted on rim of floral tube. **Ovary** superior or semi-inferior. **Fruits** capsular or loculicidal with many seeds.

LEFT TO RIGHT:
Ammannia senegalensis.
Lythrum salicaria.

Characters of similar families: Onagraceae: pollen grains usually connected by viscin threads, ovary inferior.

Herbs, annual or hardy, **sub-shrubs** or occasionally aquatics. **Stipules** absent. **Leaves** simple, opposite, alternate, or in whorls, margins entire. **Flowers** axillary, solitary or in axillary glomerules or terminal spikes, bisexual, actinomorphic or nearly so; calyx fused, contributing to a tubular hypanthium, lobes free, 8–12 parts in 2 alternating rows; corolla of 4–6 petals, sometimes clawed, inserted at the margin of the hypanthium, alternating with the divisions of the calyx; stamens 6–12 inserted within the hypanthium, sometimes dimorphic; ovary superior, style single, filiform, stigma entire, capitate sometimes bilobed. **Fruit** free, capsular, membranous, bilocular, with 4–5 locules and many ovules, tearing irregularly or loculicidal in two, sometimes more valves. **Seeds** inserted along axile placentas. **Breeding systems** can be hetero- or tri-stylous.

Literature: Graham & Graham (2014); Graham *et al.* (2005, 2011); Koehne (1903).

Lythraceae is a family of 28 genera and ca. 625 species, with several genera common in temperate zones but most species occurring in the tropics. Species of *Lythrum* are popular as ornamentals; *Lythrum salicaria* is a problematic invasive weed in North America and New Zealand. *Punica granatum* (the pomegranate) is widely cultivated.

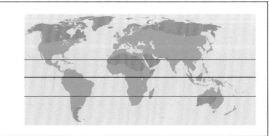

HABIT AND LEAVES

Usually herbs: **1** *Lythrum salicaria*. Leaves simple, entire, opposite and decussate: **2** *L. salicaria* and **3** *Ammannia baccifera*.

FLOWERS AND FRUIT

Hypanthium evident: **4** *Punica granatum*. Petals clawed at the base: **5** *Cuphea strigulosa*. Stamens of different lengths: **6** *Lythrum salicaria*. Ovary inferior: **7** *Lythrum hyssopifolia*. Fruits capsular: **8** *Lythrum borysthenicum*.

Onagraceae

Eve Lucas

Stipules absent
Leaves simple
Flowers bisexual
Perianth parts free
Ovary inferior

Usually **herbs**, sometimes aquatics. **Leaves** simple, margins sometimes toothed. **Flowers** showy, frequently tetramerous, hypanthium prominent; ovary inferior.

LEFT TO RIGHT:
Oenothera biennis.
Epilobium hirsutum.

Characters of similar families: Lythraceae: pollen grains free; ovary superior or semi-inferior. Scrophulariaceae (*Verbascum*): stamens 2, 4 or 5, ovary superior.

Herbs to shrubs or occasionally trees, sometimes aquatics. **Stipules** absent. Hairs simple. **Leaves** simple, alternate, opposite or whorled, margins entire or toothed, sometimes lobed, venation pinnate. **Inflorescences** terminal, axillary or solitary. **Flowers** usually bisexual, actinomorphic or zygomorphic, hypanthium usually well-developed above the ovary; (2–)4(–7)-merous; sepals free, sometimes reflexed, green or coloured as petals; petals 2–7, usually 4, free, often brightly coloured; stamens usually 8, pollen grains usually connected by viscin threads; ovary inferior, stigma capitate or 4-lobed, carpels usually 4, fused, placentation axile with 1–many ovules per locule. **Fruit** are loculidcidal capsules or berries. **Seeds** without endosperm, sometimes winged or with hairs.

Literature: Raven (1988); Wagner *et al.* (2007).

Cosmopolitan, with 21 genera and ca. 650 species of herbs or small shrubs. Native to open habitats, at altitude, in the tropics or in temperate areas. Semi-aquatics and trees also occur. Onagraceae is particularly diverse in the New World, where all genera are native, 17 exclusively so, and over ¾ of species occur.

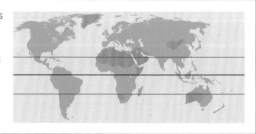

HABIT AND LEAVES

Usually herbs: **I** *Epilobium hirsutum*. Sometimes aquatics: **2** *Ludwigia sedoides*. Leaves simple, pinnate-veined: **3** *Ludwigia octovalvis*.

FLOWERS AND FRUIT

Hypanthium evident: **4** *Fuchsia triphylla*. Usually 4 showy petals: **5** *Ludwigia octovalvis*. Stamens usually 8: **6** *Epilobium angustifolium*. Stigma lobed: **7** *Epilobium hirsutum*. Ovary inferior: **8** *Circaea lutetiana*. Fruits loculicidal capsules: **9** *Epilobium hirsutum*.

Myrtaceae

Eve Lucas

Pellucid gland dots or other foliage glands
Stipules absent
Leaves simple
Leaves opposite
Ovary inferior or semi-inferior

Trees, bark peeling. **Leaves** simple, opposite with marginal veins, pellucid dots, camphor smell when crushed, margins entire. **Flowers** usually with many stamens; ovary inferior or semi-inferior. **Fruits** capsular or fleshy.

LEFT TO RIGHT:
Myrtus communis.
Eucalyptus globulus.
Decaspermum alpinum.

Characters of similar families: Clusiaceae: latex present, no aroma when crushed, primary veins parallel, ovary superior. **Oleaceae:** no aroma when crushed, leaves without pellucid glands, stamens few. **Rubiaceae:** no aroma when crushed, interpetiolar stipules distinct, leaves without pellucid glands, corolla sympetalous, stamens 4–5. **Rutaceae** (taxa with simple leaves): flowers with few stamens, ovary superior.

Trees, **treelets** or **shrubs**, bark peeling, usually mottled. **Stipules** absent. **Leaves** simple, opposite, or decussate to whorled, margins entire, lamina with pellucid gland dots, distinct intra-marginal veins and a strong smell of camphor when crushed. **Inflorescences** with flowers solitary, fasciculate, in dichasia, racemes or bottle-brush-like structures. **Flowers** usually actinomorphic; hypanthium well developed; calyx 4–5-merous, lobes free or fused, occasionally calyptrate or operculate; corolla 4–8(–12)-merous, petals free or occasionally operculate, mostly white, sometimes yellow, pink or red; stamens generally numerous, rarely 10 or fewer; ovary inferior, rarely semi-inferior, (1–)2–6(–12)-locular, (1–)2–numerous, placentation mostly axile; single style and stigma. **Fruit** woody and capsular or fleshy berries. **Seeds** mostly without endosperm, fleshy fruits with 1–2(–10) seeds per fruit, seed coat papery or hard with leafy and folded or fleshy homogenous (or distinct) cotyledons, or embryo bent or circinnate; dry fruits with numerous small, thin-walled seeds.

Literature: Brophy *et al.* (2013); EUCLID (2015); Thornhill *et al.* (2015); Wilson (2011).

Pantropical family with ca. 140 genera and ca. 5,500 species, with temperate centres of diversity in temperate Australia and at high altitudes in Southeast Asia and southern South America. The largest genus in temperate biomes is *Eucalyptus*. One species, *Myrtus communis*, commonly occurs around the Mediterranean.

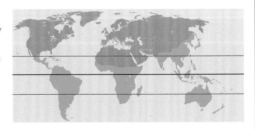

HABIT AND LEAVES

Trees with flaking bark (**1** *Angophora costata*) or shrubs (**2** *Ugni molinae*). Leaves often with pellucid gland dots: **3** *Myrtus communis*.

FLOWERS

Flowers with many stamens (**4** *Eucalyptus globulus*) or relatively few stamens (**5** *Leptospermum scoparium*). Flowers solitary: **6** *Eucalyptus gunnii*. Flowers in a bottle brush arrangement: **7** *Melaleuca linearis*.

FRUIT

Fruits capsular, free: **8** *Eucalyptus dalrympleana*. Fruits capsular, aggregated: **9** *Kunzea baxteri*. Fruits fleshy: **10** *Myrtus communis*.

Anacardiaceae

Marie Briggs

Sap or exudate present
Stipules absent
Leaves simple or compound
Leaves usually alternate
Flowers actinomorphic

Tree, shrub, occasionally herb or liana; resin canals, sap clear to milky, often drying darker. **Leaves** compound or simple, alternate or opposite, stipules absent. **Fruit** various including drupes, achenes and samaras.

LEFT TO RIGHT: *Toxicodendron wallichii*: note the pinnately compound leaves, floral nectary disk and panicle inflorescence; *Pistacia lentiscus*: note the pinnately compound leaves and separate male and female flowers; *Rhus typhina*.

Characters of similar families: **Sapindaceae:** resin absent, no resinous scent from crushed parts or bark slash. **Rutaceae:** resin absent, gland dots present, citrus or rank citrus odour from crushed parts or bark slash. **Oleaceae:** resin absent, no resinous odour from crushed parts or bark slash, with 2 or 4 stamens.

Trees, or **shrubs** (occasionally herbs or lianas). Secretory resin canals in multiple plant parts; sap clear to milky, watery or resiniferous, often drying black or brown (may cause painful irritation of the skin); fresh material often with strong turpentine or mango odour; twigs or branchlets smooth to fissured, sometimes with blotches of black dried exudate or sap. **Stipules** absent. **Leaves** simple or compound (trifoliolate or pinnate), alternate, sometimes opposite; margins entire or serrate. **Inflorescences** terminal and/or axillary (rarely cauliflorous), paniculate, racemose, spicate or thyrsoid (rarely solitary). **Flowers** small, (3–)5(–7)-merous and bisexual (occasionally unisexual and either dioecious, monoecious or polygamous); petals usually free; disk present; stamens free or rarely connate at base, 5–10, borne outside or rarely on the disk; ovary superior, rarely (semi-)inferior; 1–5-locular, always with 1 ovule per locule; style often excentric. **Fruit** various, including dehiscent or indehiscent drupes; winged fruits or fruits borne on an enlarged fleshy pedicel and receptacle.

Literature: Heywood *et al.* (2007); Johnson & More (2004); Mitchell (1990); *Plants of the World Online* (2021); Stevens (2001 onwards).

80 genera and 873 species, distributed in the tropics or subtropics, with some temperate taxa. Genera that have temperate species include *Toxicodendron* (poison oak or poison ivy), *Cotinus* (smoke-bush), *Rhus* (sumacs and varnish trees), *Schinus* (pepper trees) and *Pistacia* (turpentine tree, pistachio, and mastic). Uses include lacquer from resin (e.g. *Rhus potaninii*), edible fruits or seeds (e.g. *Pistacia vera* (pistachios)), and ornamentals (e.g. *Cotinus coggygria*).

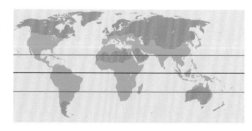

HABIT AND LEAVES

Shrubs and trees: **I** *Pistacia lentiscus*. Climbers, leaves trifoliolately compound: **2** *Toxicodendron radicans*. Leaves pinnately compound: **3** *Rhus copallinum*. Leaves simple: **4** *Cotinus coggygria*.

INFLORESCENCES

Panicles: **5** *Rhus glabra*. Spicate: **6** *Rhus aromatica*. Plants sometimes dioecious: **7** *Pistacia lentiscus* (male flowers) and **8** *P. lentiscus* (female flowers).

FRUITS

Brightly coloured drupes with slightly excentric styles: **9** *Pistacia lentiscus*, **10** *Rhus copallinum* and **11** *Pistacia terebinthus*.

Sapindaceae

Marie Briggs

Stipules absent
Leaves compound or simple
Leaves alternate or opposite
Flowers unisexual
Ovary superior

Trees or **shrubs** (rarely lianas or herbs). **Stipules** usually absent. **Leaves** simple or compound, opposite or alternate. **Flowers** zygomorphic or actinomorphic, often unisexual, often with 8 stamens. **Fruit** often lobed or winged.

LEFT TO RIGHT:
Acer diabolicum: note fascicles of flowers, separate male and female flowers, and the samara fruits;
Acer saccharinum: note young simple leaves and samara fruits.

Characters of similar families: Rutaceae: gland dots present in leaves and other plant parts, citrus or rank citrus odour from crushed parts or bark slash. **Anacardiaceae:** sap from resin canals, mango or resinous odour from crushed parts or bark slash. **Juglandaceae:** often aromatic, leaflets with peltate scales on underside, basal leaflets often smaller than terminal ones, flowers small, often in catkins or spikes. **Oleaceae:** flowers usually bisexual, sweetly scented, with 2 or 4 stamens.

Trees, shrubs or **lianas**, rarely **herbs**; sap from excretory canals absent but xylem-derived sap sometimes present (e.g. in *Acer* species), clear, not turning black; fresh material usually without strong odour when crushed. **Stipules** usually absent (pseudo-stipules or minute stipules occasionally present). **Leaves** alternate or opposite, variously compound (pinnate, bipinnate, palmate, trifoliolate or unifoliolate) or simple, petiole bases often swollen. **Inflorescence** terminal or axillary, less commonly cauliflorous, paniculate, racemose or thyrsoid, fascicled in leaf axils or solitary. **Flowers** actinomorphic, occasionally zygomorphic (e.g. in *Aesculus* species), usually functionally unisexual, sometimes bisexual; plants andromonoecious, androdioecious, or dioecious; sepals (3–)4–5(–8), fused or free, petals (0–)4–5(–6) free, often clawed when present; disk present or reduced to a pair of glands; stamens free, 4–14(–74) but most commonly a single whorl of 8, usually inserted inside the disk, sometimes on the disk surface, filaments often hairy; ovary superior, 1–3(–8)-celled, lobed or not. **Fruit** drupes, berries, capsules, samaras or samaroid, often 2–3-lobed or winged, may be composed of 1 or 2 mature carpel(s), with aborted vestigial carpels sometimes present.

Literature: Heywood *et al.* (2007); Johnson & More (2004); Plants of the World Online (2021); Stevens (2001 onwards); Wu *et al.* (2007).

A cosmopolitan family of 144 genera and 1925 species. Now including the largely temperate families Hippocastanaceae (horse chestnuts or buckeyes) and Aceraceae (maples). Many *Acer* species are valued for their beautiful foliage and bark; maple syrup is made from the xylem sap of other species (e.g. *Acer saccharinum*, the sugar maple).

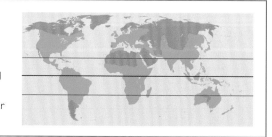

HABIT AND LEAVES

Leaves simple: **1** *Acer platanoides*. Leaves palmately compound: **2** *Aesculus hippocastanum*. Leaves pinnately compound: **3** *Alectryon excelsus* (see also **14** *Dipteronia sinensis*).

INFLORESCENCES

Inflorescences: **4** *Aesculus indica* and **5** *Acer rubrum*

FLOWERS

Flowers zygomorphic: **6** *Koelreuteria paniculata* (note hairy filaments) and **7** *Aesculus parviflora*. Actinomorphic: **8** *Acer pictum* (note fleshy disk) and **9**, **10** *Alectryon excelsus* (male and female).

FRUIT

Capsules: **11** *Koelreuteria bipinnata* (inflated), **12** *Aesculus hippocastanum* (fleshy) and **13** *Alectryon excelsus* (note fleshy arillode at base of seed). Samaroid: **14** *Dipteronia sinensis*.

133

Rutaceae

Marie Briggs

Oily gland dots
Leaves often compound
Leaves alternate or opposite
Flowers usually bisexual
Ovary superior

Trees, **shrubs**, **herbs** or **lianas**. Crushed fresh **leaves** with citrus or rank citrus scent. **Sap** absent. **Stipules** absent. **Leaves** compound, simple or unifoliolate, alternate or opposite; oily gland dots in **leaves**, **flowers** and **fruit**.

LEFT TO RIGHT:

Haplophyllum acutifolium: note simple leaves, superior ovary with oily gland dots, on top of a fleshy disk;

Citrus × limon: note fleshy disk at base of ovary and section through fruit, showing seeds;

Boenninghausenia albiflora: a woody herb with tripinnately compound leaves.

Characters of similar families: Anacardiaceae: oily gland dots absent, style excentric, mango or resinous scent to crushed parts, sap present. **Sapindaceae:** oily gland dots absent, filaments often hairy, free rachis tip in pinnately-compound leaves, no strong scent in crushed parts. **Juglandaceae:** oily gland dots absent, leaflets with peltate scales on underside, flowers small, often in catkins or spikes. **Myrtaceae:** ovary inferior, many stamens, leaves simple, usually opposite, leaf margins entire, crushed parts usually scented but not citrus or rank citrus.

Trees, **shrubs**, some **lianas** and **herbs**, occasionally scandent; thorns or spines sometimes present; oil glands present, often visible on leaves, flowers and fruit, ± regular, crushed parts with citrus or rank citrus scent. **Sap** absent. **Stipules** absent. **Leaves** pinnately (once or more), palmately or trifoliolately compound, sometimes simple or unifoliolate; alternate, opposite or whorled; rachis with or without wings. **Inflorescences** variable including cymes, racemes, and panicles, sometimes reduced to single flower, terminal and/or axillary. **Flowers** usually bisexual, sometimes unisexual (monoecious or polygamomonoecious), 3–5-merous; stamens 2–many, arranged in 2 whorls, often in a ring, filaments sometimes thick and fleshy, sometimes flattened, connate at the base or not; disk conspicuous, ovary usually superior, with (1–) 4–5(–to many) locules. **Fruit** variable, including hesperidiums (e.g. *Citrus*), berries (e.g. *Glycosmis*), capsules (e.g. *Ruta*), winged (*Ptelea*) and drupaceous (e.g. *Skimmia*). **Seeds** sometimes black and shiny.

Literature: Heywood *et al.* (2007); Johnson & More (2004); Plants of the World Online (2021); Stevens (2001 onwards).

161 genera and 2085 species. Mainly tropical and subtropical with a few genera extending into temperate regions. Known for edible fruits and an abundance of essential oils. Members of the genus *Citrus* are naturalised in some warmer temperate zones, where they are extensively cultivated. Other taxa (e.g. *Skimmia* and *Choisya*) are common ornamentals.

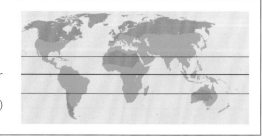

LEAVES

Unifoliolate leaves: **1** *Citrus* sp. (note gland dots). Trifoliolately compound leaves: **2** *Melicope simplex*. Pinnately compound leaves: see **5**. Spines present in some species: **3** *Citrus trifoliata*.

INFLORESCENCES AND FLOWERS

Inflorescences: include cymes **4** *Leionema nudum* and racemes **5** *Dictamnus albus* var. *purpureus*. Flowers 3–5-merous: **6** *Cneorum tricoccon*, **7** *Melicope simplex*, **8** *Correa backhouseana*, **9** *Ruta angustifolia* (note gland dots on ovary) and **10** *Citrus trifoliata*.

FRUIT

Various fruits including hesperidiums: **11** *Citrus trifoliata*. Berries: **12** *Glycosmis pentaphylla*. Samaras: **13** *Ptelea trifoliata*. Capsules: **14**, **15** *Melicope ternata* (note shiny black seeds).

Malvaceae–Malvoideae

Sara Edwards, Sue Frisby and Nicky Biggs

Stipules present
Leaves simple
Leaves alternate
Flowers bisexual
Flowers actinomorphic

Herbs, **shrubs** or **trees**. **Leaves** spirally arranged, 3,5,7-veined from base, simple, or dissected. **Indumentum** stellate, simple and/or glandular. **Epicalyx** present or absent; stamens fused into a tube. Fruit a schizocarp or loculicidal capsule.

LEFT TO RIGHT: *Althaea rosea*; *Pavonia senegalensis*; *Malva sylvestris*.

Characters of similar families: Malvaceae–Byttnerioideae: leaves usually unlobed, epicalyx usually present, rarely absent, staminodes usually present, anthers 2- or 3- thecate, style simple. Bixaceae (*Cochlospermum* spp.): trees or shrubs, leaves palmately-lobed, filaments free, anthers dehisce via pores, fruit obovoid or ellipsoid, 3–5-valved. Passifloraceae–Turneroideae (especially *Piriqueta* spp.): epicalyx absent, stamens 5 free, styles 3, fruit a 3-valved capsule, dehiscence loculicidal. Cistaceae: leaves pinnately veined, petals and sepals free, filaments distinct.

Herbs, **shrubs** and **trees**. **Indumentum** usually stellate, simple or glandular; sometimes with extra-floral nectaries; inner bark tough-fibrous. **Stipules** present, often falling early. **Leaves** alternate or spirally arranged; simple or lobed, ovate, lanceolate or cordate; venation palmate or 3,5,7-veined from base; margins entire, crenate, cordate or serrate. **Inflorescence** axillary or terminal, solitary or fasciculate panicles, racemes or spikes. **Flowers** usually bisexual, actinomorphic; epicalyx present or absent, of 2–many bracts; calyx tubular, 5-lobed, fused basally; petals 5, often clawed, adnate to base of staminal tube; stamens few to many, staminodes absent or relatively small, filaments fused into a tube, anthers usually monothecal; ovary superior, 1–40-locular, styles usually branched. **Fruit** a capsule or schizocarp with variously ornamented mericarps, or rarely a berry. **Seeds** solitary to numerous, woolly pubescent to glabrous; endosperm oily.

Literature: Bayer & Kubitzki (2002); Cheek (2007); Fryxell (1997); Hanes (2015).

Malvoideae has ca. 59 genera and ca. 734 species, in temperate regions of the Americas, Europe, Asia, Australasia and South Africa; a further ca. 1,500 species are primarily tropical or subtropical. Malvoideae is now a sub-family of Malvaceae. Major genera include *Alcea*, *Malva*, *Lavatera*, *Hibiscus* and *Pavonia*. *Gossypium* is used for fibres, *Hibiscus* for ropes and teas, and *Hibiscus*, *Abutilon* and *Lavatera* as ornamentals.

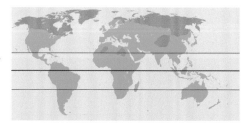

HABIT

Herbs, shrubs or trees: **1** *Malvastrum americanum*, **2** *Cienfuegosia argentina* and **3** *Lavatera* spp. **4** Stellate hairs.

FLOWERS

Staminal tube: **5** *Hibiscus moscheutos*, **6** *Sidastrum paniculatum* and **7** *Abutilon austroafricanum*. Epicalyx: **8** *Hibiscus sororius*. Epicalyx red: **9** *Pavonia* × *gledhillii*.

FLOWERS

Capsules or schizocarps: **10** *Hibiscus micranthus*. Enclosed by calyx, surrounded by epicalyx: **11** *Pavonia burchellii*. Schizocarp: **12** *Abutilon austroafricanum*. Capsule: **13** *Gossypium barbadense*.

137

Thymelaeaceae

Laura Jennings

Stipules absent
Leaves always entire
Flowers actinomorphic
Perianth parts fused
Ovary superior

Usually **shrubs** (may be trees, herbs or lianas), bark fibrous, stipules absent. **Leaves** simple and entire. **Flowers** tubular; perianth hairy, in a single whorl; ovary superior. **Fruit** usually a drupe or nutlet.

LEFT TO RIGHT:
Daphne giraldii;
Stellera chamaejasme.

Characters of similar families: Clusiaceae: yellow sap present, flowers unisexual with separate calyx and corolla, petals free. **Ebenaceae:** stipules present, leaves with glands at base or on petiole, flowers unisexual with separate calyx and corolla. **Lamiaceae:** flowers zygomorphic with separate calyx and corolla, stamens usually 2. **Rubiaceae:** stipules present, flowers with separate calyx and corolla, ovary inferior. **Cistaceae:** calyx not coloured and petaloid, sepals and petals free.

Commonly **shrubs**, also tall **trees**, **lianas**, or rarely **herbs**, stems flexible with silky fibres present, often visible at the end of cut stems. **Stipules** absent. **Leaves** simple, entire, alternate or opposite, petiolate or sessile, often clustered at the end of the branches. **Inflorescences** axillary or terminal, usually umbelliform, spikes or heads. **Flowers** actinomorphic, bisexual, pedicels often articulated; perianth tubular, in a single whorl with 4–5 lobes at the rim, often brightly coloured, often hairy; stamens inserted inside the rim of the tube (free in *Tepuianthus*), ovary superior, usually 1 ovule per locule. **Fruit**, often fleshy and indehiscent or capsular. **Seeds** 1 per locule, sometimes with an appendage or aril.

Literature: Beaumont *et al.* (2009); Ding Hou (1960); Herber (2003); Rogers (2009); Wurdack & Horn (2001).

Thymelaeaceae is a widely distributed family of 50 genera and ca. 900 species. The genus *Tepuianthus* (previously in its own family) was added in 2001. The genus *Gnidia* was found to be polyphyletic but not all species in the clade have been reclassified. Used as agarwood (*Aquilaria*) and ramin (*Gonystylus*).

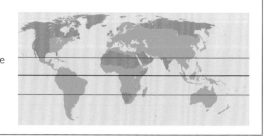

HABIT

Shrubs: **1** *Gnidia oppositifolia*. Large trees: **2** *Gonystylus confusus*. Rarely herbs: **3** *Kelleria dieffenbachii*. Fibres visible at end of cut twigs: **4** *Aquilaria* sp.

LEAVES

Leaves opposite: **5** *Pimelea linifolia*. Leaves alternate: **6** *Gyrinops walla*. Leaves clustered at end of branches: **7** *Thymelaea tarton-raira*.

FLOWERS

Flowers with 1 perianth whorl: **8** *Daphne retusa*. Stamens attached to inside of tube: **9** *Edgeworthia chrysantha*.

FRUIT

Fruit drupaceous: **10** *Daphne mucronata* ssp. *linearifolia* Fruit capsular: **11** *Gyrinops walla*. Seeds ornamented: **12** *Aquilaria malaccensis*.

Cistaceae

Tony Hall

Leaves simple
Leaves alternate or opposite
Flowers actinomorphic
Perianth parts free
Ovary superior

Shrubs, **subshrubs** or **herbs**. **Leaves** alternate or opposite, simple, entire with 3 more-or-less parallel veins. **Inflorescence** solitary or in cymose inflorescences. Stamens numerous. Style simple. **Fruit** a loculicidal capsule.

LEFT TO RIGHT:
Tuberaria globulariifolia;
Lechea lakelae.

Characters of similar families: **Malvaceae:** leaves commonly lobed, calyx tubular, stamens joined at their bases, often forming a tube around the pistils. **Thymelaeaceae:** stems with silky fibres present, calyx and hypanthium coloured and petaloid. **Bixaceae:** inflorescence a terminal panicle, fruit a capsule, loculicidally 2-valved, usually spiny.

Shrubs, subshrubs, some annual **herbs**, aromatic, often with stellate or branched hairs or peltate scales. **Stipules** present or absent. **Leaves** simple, opposite or alternate and spirally arranged, margins entire, occasionally wavy, usually pinnately veined. **Inflorescences** terminal or axillary, solitary or cymose, 1–10. **Flowers** yellow, white, pink to reddish or purple, occasionally blotched; bisexual, actinomorphic, hypogynous, chasmogamous or cleistogamous (e.g. *Lechea*). **Sepals** 3 or (4)5, equal or subequal; if 5, two are often much narrower and sometimes shorter. Both sepals and petals are free. **Petals** usually 5(3) or 0 in cleistogamous flowers; ovary superior, usually with 1–3 locules in each carpel; style 1, simple; stamens usually numerous, filaments distinct; anthers 2-locular, pollen grains usually tricolporate. **Fruit** non-fleshy, dehiscent; loculicidal capsule or valvular with 3, 5 or up to 11 valves, opening from the top down. **Seeds** endospermic, usually small, wingless, dark brown or black.

Literature: Heywood (2007b); Nandi (1998); Watson & Dallwitz (1992 onwards).

Widespread, with eight genera and ca. 200 species, the majority in temperate regions with the highest concentration in the Mediterranean region, especially the Iberian Peninsula: *Cistus*, *Fumana*, *Helianthemum* and *Tuberaria*. *Hudsonia* occurs in temperate North America. Four genera include species that are native to the neotropics: *Crocanthemum*, *Helianthemum*, *Lechea* and *Pakaraimaea*.

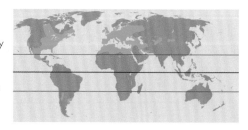

HABIT

Shrubs or subshrubs: **1** *Helianthemum alypoides* and **2** *Cistus atriplicifolius*. Annual: **3** *Tuberaria guttata*.

LEAVES AND INFLORESCENCES

Leaves simple, opposite: **4** *Cistus albidus*. Cymose inflorescence: **5** *Cistus clusii*.

FLOWERS AND FRUIT

Flowers actinomorphic, petals free: **6** *Cistus halimifolius*. Ovary superior, stamens numerous: **7** *Cistus salviifolius*. Petals with basal blotch: **8** *Cistus ladanifer*. Fruit a capsule: **9** *Cistus laurifolius* and **10** *Tuberaria lignosa*.

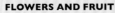

Brassicaceae

Carmen Puglisi

Stipules absent
Leaves simple or compound
Leaves alternate
Flowers bisexual
Ovary superior

Leaves alternate, mustard smell. **Flower** cruciferous, with 4 sepals, 4 petals, 6 stamens and superior ovary. **Fruit** usually a siliqua, a bivalved capsule with persistent replum.

LEFT TO RIGHT:
Brassica spinescens;
Menonvillea nordenskjoeldii;
Tomostima platycarpa.

Characters of similar families: Capparaceae: woody, stamens often numerous and long exserted, capsules lacking replum. **Cleomaceae:** leaves compound, stipules present. **Leguminosae:** leaves usually compound, stipules present, flowers usually 5-merous. **Caryophyllaceae:** leaves usually opposite, inflorescences dichasial, flowers usually 5-merous.

Herbs or **subshrubs**, annual or perennial, occasionally succulent, some species aquatic; stems often hollow along internodes; often producing mustard oils. **Stipules** absent. **Leaves** simple, rarely pinnately compound, alternate, rarely opposite, herbaceous or rarely fleshy, petiolate or sessile; lamina entire, lobed or dissected; indumentum often present. **Inflorescences** racemes or corymbs, rarely spikes or a solitary flower, often ebracteate, axillary, terminal or leaf-opposed. **Flowers** bisexual, cruciferous (cross-shaped); sepals 4, usually free, in two decussate whorls (K2+2); petals 4, rarely absent, free, usually clawed; stamens usually 6, rarely 4, 2, or many, 2 (often smaller) stamens in the outer, alternipetalous whorl, 4 in the inner, alternisepalous whorl, these sometimes fused in pairs; filaments sometimes appendaged or winged; anthers basifixed, 2–(1)-locular, dehiscing longitudinally; nectary often present, annular or divided, hypogynous; ovary superior, carpels 2, fused, secondarily divided by a false septum (replum), ovules 1–many, placentation parietal; style 1, sometimes reduced, stigmas 1–2, often lobed. **Fruit** often a bilocular capsule dehiscing in 2 valves (a siliqua or a silicula, according to length/width ratio), sometimes indehiscent, schizocarpic, lomentaceous or samaroid. **Seeds** medium-small, often mucilaginous, without endosperm.

Literature: Allan (1961); Appel & Al-Shehbaz (2003); Hedge (1968); Koch *et al.* (2018); Watson & Dallwitz (1992 onwards).

Brassicaceae has 351 genera and ca. 3,977 species; nearly cosmopolitan, most diverse in temperate Eurasia. The largest genus is *Draba* (ca. 400 species). Several genera (including *Brassica*, *Raphanus* and *Rorippa*) are important food crops, others (such as *Arabis* and *Lobularia*) are common ornamentals. The family's tribes and genera have been re-circumscribed in recent years as a result of phylogenetic research.

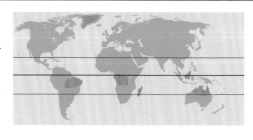

HABIT AND EXCEPTIONAL FEATURES

Habit: **1** *Lobularia maritima* and **2** *Notothlaspi rosulatum*. Compound leaves: **3** *Cardamine macrophylla*. Adapted to aquatic (**4** *Rorippa palustris*) and arid (**5** *Zilla macroptera*) habitats.

FLOWERS

Typical flower structure from above (**6** *Pachycladon exile* and **7** *Heliophila coronopifolia*) and the side (**8** *Raphanus raphanistrum*). Apetalous flower: **9** *Cardamine verna*.

FRUIT

Racemose inflorescence with developing siliquas: **10** *Brassica fruticulosa*. Siliqua: **11** *Enarthrocarpus arcuatus* and **12** *Enarthrocarpus pterocarpus*. Silicula: **13** *Biscutella didyma*. Dehisced fruits with persistent replum: **14** *Lobularia maritima*. Seeds: **15** *Cardamine unguiculus*.

Plumbaginaceae

Sue Frisby

Stipules absent
Leaves simple
Leaves alternate
Flowers actinomorphic
Ovary superior

Leaves spirally arranged, often a basal rosette, sometimes auriculate; secretory glands exuding water, salts or mucilage common. **Inflorescence** bracteate, usually terminal. **Flowers** bisexual, tubular, pentamerous; ovary superior, one small seed.

LEFT TO RIGHT: *Plumbago amplexicaulis*; *Limonium spectabile*; *Armeria maritima*; *Plumbago auriculata*.

Characters of similar families: Polygonaceae: swollen nodes, stipules present, ochreas present. **Primulaceae:** secretory glands absent, sepals not usually colourful, showy or paper-like, seeds several to many. **Droseraceae:** insectivorous, enzyme-secreting leaves, intrapetiolar stipules often present, seeds 3 to numerous. **Simmondsiaceae:** dioecious, leaves opposite, flowers apetalous. **Frankeniaceae:** leaves opposite, revolute, clawed petals, carpels 1–3.

Herbs, small caespitose **shrubs**, or **lianas**. Secretory glands that exude water, salt or mucilage often present. **Hairs** simple or lepidote indumentum sometimes present. **Stipules** absent. **Leaves** simple, entire to lobed, alternate, sometimes auriculate or in basal rosettes, petiolate to sessile, epulvinate, membranous or leathery, pinnately veined. **Inflorescences** terminal or axillary cymes, racemes, panicles, corymbs or solitary heads, herbaceous bracts scarious, sometimes absent. **Flowers** bisexual, actinomorphic, 5-merous, sepals may be colourful and showy, often thin and papery, connate, tube 5–10-lobed, sometimes with spiky glandular trichomes (*Plumbago*), petals free or fused into a long tube (*Plumbago*), imbricate, clawed, often persistent; stamens 5, opposite petals, mostly free (*Plumbago*) or epipetalous at base of corolla (*Limonium*); anthers 2-locular dehiscent longitudinally; styles 1, or 5, opposite sepals, free to connate; ovary superior, of 5 fused carpels, unilocular, basal placentation, ovule 1. **Fruit** a capsule or achene with circumscissile or irregular dehiscence, scarious bracts, bracteoles or glandular calyx for wind dispersal, or a spongy mesocarp for water or ocean dispersal. **Seeds** with or without endosperm, small, often winged.

Literature: Culham (2007c); Koutroumpa *et al.* (2018); Kubitzki (1993d); Tutin *et al.* (1972a; Watson & Dallwitz (1992 onwards).

Cosmopolitan, frequent in coastal, dry, cold, saline regions: 25 genera and ca. 600 species in two subfamilies. Plumbaginoideae, such as *Plumbago* (leadwort), are predominantly warm temperate. Limonoideae are predominantly cooler temperate, with *Armeria* and *Limonium* the largest genera. *Plumbago*, *Armeria*, *Limonium* and *Ceratostigma* are horticulturally valuable; *Plumbago* and *Limonium* are used medicinally.

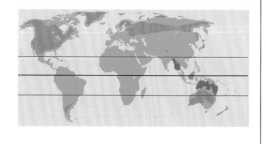

HABIT
Frequently coastal or saline habitat, often herbs: **1** *Armeria maritima*. Small caespitose shrubs: **2** *Acantholimon litvinovii*. Lianas, simple leaves: **3** *Plumbago auriculata*.

INFLORESCENCES AND FLOWERS
Cymes: **4** *Limonium arborescens*. Branched spikes: **5** *Psylliostachys spicatus*. Solitary heads: **6** *Armeria maritima*. Scarious bracts, showy sepals: **7** *Limonium sinuatum*. Glandular trichomes on connate sepals, long corolla tube: **8** *Plumbago zeylanica*. Stamens opposite petals: **9** *Armeria maritima*.

FRUIT AND SEEDS
Fruits encased in scarious calyx: **10** *Armeria maritima* and **11** *Armeria alliacea*.

145

Polygonaceae

Nicky Biggs

Stipules present
Leaves simple
Leaves usually alternate
Flowers usually bisexual
Ovary superior

Mostly **herbs**, stems with swollen nodes and characteristic ocrea (not Eriogonoideae). **Leaves** alternate. **Flowers** small and subtended by involucral bracts. **Fruits** achene-like, or a dry trigonous or flattened nut.

LEFT TO RIGHT:
Bistorta amplexicaulis;
Persicaria lapathifolia: ocrea conspicuous at leaf nodes, in fruit;
Persicaria amphibia: in flower.

Characters of similar families: Caryophyllaceae: ocrea absent, leaves opposite, petals usually present (0–1)4–5. Plumbaginaceae: nodes swollen but ocrea absent, stipules absent.

Trees, **shrubs**, **vines** or **herbs**, latex absent. **Stems** prostrate to erect, nodes often swollen, rarely tendrils. Stipules present, rarely absent, often with a nodal sheath (ocrea) (Polygonoideae), cylindric to funnelform, chartaceous, membranous or foliaceous. **Leaves** simple, basal, or basal and cauline, forming rosette, mostly alternate (some sub-opposite), margins mostly entire, occasionally lobed. **Inflorescences** terminal or axillary, spicate, racemose, paniculate or solitary, bracts absent or 2–10, foliaceous or scale-like, peduncle present or absent, flowers subtended by involucral bracts or enclosed in tubular involucres or subtended by connate bracteoles forming a persistent membranous tube (ocreola). **Flowers** small, actinomorphic, usually bisexual, 1–many; perianth persistent, often accrescent in fruit, green, white, yellow, red, or purple, campanulate to urceolate, sometimes membranous, or fleshy, glabrous or pubescent sometimes glandular; tepals (3)5–6 free or forming a tube, usually in 2 whorls; stamens (1)6–9(–18), staminodes rarely present, filaments free or forming a tube; ovary superior, carpels 2–4 united to form 1 locule, styles 1–3(4), stigmas 1 per style. **Fruits** achenes, trigonous or lenticular nut, winged or unwinged, glabrous or pubescent, yellow, brown, red, or black. **Seeds** 1.

Literature: Freeman & Reveal (2005); Heywood (1993); Li *et al.* (2003).

Polygonaceae has 55 genera, ca.1,000 species in two subfamilies, Polygonoideae and Eriogonoideae. The family is cosmopolitan and common in northern temperate regions. The largest genera are *Polygonum* (ca. 250 species), *Rumex* (ca. 200 species), *Eriogonum* (ca. 200 species), *Persicaria* (ca. 100 species) and *Calligonum* (ca. 80 species). Various species are edible, for example the petioles of *Rheum* (rhubarb) and seed of *Fagopyrum* (buckwheat); some have medicinal use or are ornamentals.

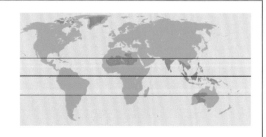

HABIT AND LEAVES

Usually shrubs, herbs: **1** *Rheum officinale*, **2** *Rumex cyprius* and **3** *Rumex vesicarius*. Vines: **4** *Fallopia baldschuanica*. Sometimes fleshy: **5** *Polygonum maritimum*. Leaves alternate: **6** *Rumex sanguineus*. Ocrea present: **7** *Persicaria microcephala* 'Red Dragon' and **8** *Persicaria longiseta* var. *rotundata*.

INFLORESCENCES AND FLOWERS

Flowers: racemose (**9** *Bistorta amplexicaulis* and **10** *Bistorta affinis*), paniculate (**11** *Fagopyrum esculentum*) or spicate (**12** *Rheum officinale*). Flowers bisexual: ovary superior, tepals 4 (**13** *Atraphaxis pyrifolia*), tepals commonly 6, free, in 2 whorls (**14** *Eriogonum umbellatum* and **15** *B. affinis*).

FRUIT

Fruit a lenticular nut: **16** *Rumex cassius*. Achenes trigonous and/or winged to aid dispersal by wind: **17** *Rumex cyprius* and **18** *Oxyria digyna*.

Droseraceae

Martin Cheek

Stipules present
Leaves simple
Leaves alternate
Leaves entire
Ovary superior

Leaves circinnate, alternate or spiral, often in sessile rosettes, rarely in verticils, covered in sessile and stalked glandular hairs with red centres in the middle of mucilage globules. (*Dionaea* and *Aldrovanda* lack hairs but the two halves of the leaf blade close swiftly when triggered).

LEFT TO RIGHT:
Drosera capensis;
Drosera prostratoscaposa.

Characters of similar families: Byblidaceae (in New Guinea and Australia only): stipules absent, stalked glands of leaves entirely colourless (lacking red head), glands immobile, leaves not circinnate. **Roridulaceae** (South Africa only): stipules absent, leaf and stem glands immobile, leaves not circinnate. **Drosophyllaceae** (in the western Mediterranean): shrubby, leaves reverse circinnate, glandular hairs immobile.

Carnivorous **herbs**, short-lived or perennial, with underground rootstocks, forming sessile rosettes or aerial stems, rarely verticillate and/or twining (Australia). **Sap** absent. **Hairs** glandular conspicuous and dense on the adaxial leaf-blades, short and immobile and long and mobile, attracting and trapping prey; stalked, the head usually red, forming a sphere of mucilage (hairs absent in *Dionaea* and *Aldrovanda*, which have leaf-blades that are mechanical snap-traps instead). **Stipules** adaxial. **Leaves** simple, circinnate, alternate or spiral, margins entire. **Inflorescences** appearing axillary, or terminal, usually racemoid cymes, rarely subpaniculate or single-flowered. **Flowers** bisexual, actinomorphic, small or large, white, red or yellow; perianth 5-merous, sepals 5, free; petals 5, free; stamens usually 5(4–20) free, 2–4-locular, dehiscence longitudinal, pollen in tetrads; ovary superior, unilocular, ovules numerous, styles 3–5 sometimes branched. **Fruit** dry, dehiscent capsule; seeds numerous, minute.

Literature: Lowrie (1987–1998).

A temperate and tropical family of 3 genera and about 250 species. *Dionaea* (SE USA), the Venus flytrap, is terrestrial and has mechanical traps, as does the aquatic, free-floating, rootless widespread and threatened Old World (Eurasia to Australia) *Aldrovanda* (both of these genera are monotypic). *Drosera* is most diverse in Brazil and SW Australia, with a secondary centre in the Cape of South Africa. In temperate areas, the species usually occur in nutrient-poor bogs.

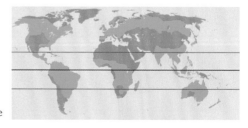

HABIT AND LEAVES

Sessile rosettes: **1** *Drosera intermedia* and **2** *Drosera adelae*. Glandular mucilaginous hairs: **3** *Drosera rotundifolia* and **4** *Drosera auriculata*. Snap traps (glandular hairs absent): **5** *Dionaea muscipula*.

FLOWERS

Bisexual, actinomorphic, 5-merous. **6** *Dionaea muscipula*. Style branched: **7** *Drosera rotundifolia* and **8** *Drosera uniflora*.

Caryophyllaceae

Saba Rokni

Leaves simple
Leaves usually opposite
Leaves entire
Flowers actinomorphic
Ovary superior

Mostly **herbs**. Stems often swollen at nodes. **Leaves** simple, entire, opposite, often connate. **Flowers** actinomorphic, petals free, often clawed. **Fruit** usually a many-seeded capsule dehiscing at apex by teeth or valves.

LEFT TO RIGHT:
Dianthus leucophoeniceus;
Silene dioica.

Characters of similar families: Amaranthaceae: stipules absent, leaves not connate, margins may be lobed or dentate, perianth uniseriate, capsules circumscissile, betalain pigments present. Gentianaceae: corolla tubular, stamens adnate to the corolla tube, fruit a 2-valved septicidal capsule.

Herbs or **sub-shrubs**, rarely **shrubs** or **small trees**. **Stems** often swollen at nodes. Stipules occasionally present, mostly scarious. **Leaves** simple, opposite decussate, rarely alternate or whorled, often connate, entire, sometimes succulent. **Inflorescences** cymose, shape highly variable, rarely single-flowered. **Flowers** actinomorphic, bisexual, rarely unisexual; sepals usually (4–)5 and leafy or scarious, free or fused into a tube, persistent, mostly subtended by bracts; petals (0–)4–5, rarely more, free, often clawed; limb entire, bifid, lacerate, or variously divided, coronal scales sometimes present at juncture with claw; staminodes sometimes present; stamens 1–2x the sepal number, rarely fewer, in 1 or 2 whorls; filaments free or connate at base; ovary superior, of 2–5(–10) united carpels, 1-locular, sometimes basally imperfectly 2–5-locular, ovules (1–)many; styles (1–)2–5(–6), usually filiform, sometimes united at base; petals, stamens and ovary sometimes separated from calyx by an anthophore. **Fruit** a capsule dehiscing at apex by teeth or valves, rarely an achene or berry. **Seeds** small, (1–)many, usually globose to pyriform or reniform and laterally compressed, testa sculptured, rarely smooth, embryo usually curved around perisperm.

Literature: Bittrich (1993); Lu *et al.* (2001); Heywood *et al.* (2007); Rabeler & Hartman (2005); Walters (1993).

About 100 genera and 2,200–3,000 species: mainly distributed in the temperate regions of the northern hemisphere, the centre of diversity being the Mediterranean and Irano-Turanian regions. *Silene* is the largest genus, other large genera include *Dianthus*, *Arenaria*, *Cerastium*, *Gypsophila*, *Stellaria* and *Paronychia*. Several species are cultivated as ornamental flowers; some are widespread annual weeds.

HABIT, LEAVES AND STEMS

Herbaceous, leaves simple, opposite, entire: **I** *Stellaria neglecta*. Leaves connate: **2** *Cerastium arvense*. Nodes swollen: **3** *Acanthophyllum gypsophiloides*. Succulent leaves and scarious stipules: **4** *Spergularia rubra* and **5** *Paronychia* sp.

FLOWERS

Calyx tubular: **6** *Silene korshinskyi*. Petal clawed with coronal scales: **7** *Silene boryi*. Petals bifid, flower 4-merous: **8** *Cerastium diffusum*. Lacerate petals and filiform styles: **9** *Dianthus deltoides*. Petals entire, stamens in 2 whorls: **10** *Spergularia rubra*. Petals deeply divided: **II** *Silene flos-cuculi*. Bracts silvery-scarious, petals absent, staminodes present: **12** *Paronychia chionaea*. Bracts brown-scarious and enclosing inflorescence: **13** *Petrorhagia nanteuilii*.

FRUIT AND SEEDS

Capsules: **14** *Cerastium fontanum*, **15** *Silene latifolia* and **16** *Sagina procumbens*. Berry: **17** *Silene baccifera*. Seeds: **18** *Gypsophila cephalotes* and **19** *Dianthus recticaulis* (dissected capsule).

Amaranthaceae

Saba Rokni

Stipules absent
Leaves simple
Leaves alternate or opposite
Flowers actinomorphic
Ovary superior

Usually **herbs** and **shrubs**. **Leaves** simple, sometimes reduced or succulent. **Flowers** uniseriate, usually with variously shaped bracts and bracteoles. **Fruit** an achene or capsule, rarely a berry. **Seeds** 1(–many).

LEFT TO RIGHT: *Beta vulgaris*; *Suaeda maritima*; *Amaranthus retroflexus*.

Characters of similar families: Caryophyllaceae: stipules may be present, leaves connate, flowers often have a clearly distinguishable calyx and corolla, petals free and often clawed, capsules dehiscing at apex by teeth or valves. **Polygonaceae:** stems with characteristic ocrea.

Herbs and **shrubs**, rarely small **trees** or **lianas**. **Stems** frequently angular or grooved, sometimes succulent and apparently jointed, nodes often swollen. **Stipules** absent. **Leaves** alternate or opposite, simple, sometimes reduced or succulent, indumentum sometimes mealy, margins entire or lobed or toothed. **Inflorescences** cymes, spike or head-like or in glomerules, or flowers solitary, terminal or axillary. **Flowers** generally small, mostly actinomorphic, unisexual or bisexual; usually with variously shaped (sometimes showy) **bracts** and **bracteoles**; lateral flowers occasionally sterile, modified into scales, spines, bristles or hairs; perianth uniseriate, herbaceous, succulent, membranous, or papery; tepals (0–)5(–8), free or fused, variously coloured and may persist in fruit, becoming hardened or enlarged and modified; stamens fewer than 5 or isomerous with tepals, rarely more, filaments free or fused; ovary superior, seldom semi-inferior, unilocular, ovules 1(–many); stigmas 1–3(–6), elongate to capitate. **Fruit** an achene or capsule with irregular or circumscissile dehiscence, or indehiscent, rarely a berry; usually subtended or enclosed by persistent (sometimes enlarged and modified) perianth or bracts or bracteoles. **Seeds** 1(–many), globular, lenticular, or ovoid, morphology variable; embryo curved to spiral.

Literature: Bao *et al.* (2003); Crawford (2015a); Zhu *et al.* (2003); Kühn *et al.* (1993); Townsend (1993).

Worldwide with 186 genera and 2,050–2,500 species. Amaranthaceae in the broad sense currently includes Chenopodiaceae. A morphologically variable family divided into several sub-families. The family includes many drought and salt-tolerant species and several widespread invasive weeds. Several species have been domesticated as food crops, including beetroot, spinach, quinoa, and American grain amaranths.

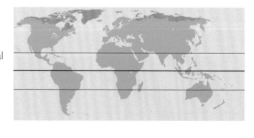

HABIT AND HABITAT

Usually herbs and shrubs, often ruderal or in deserts and saline habitats: **1** *Amaranthus retroflexus*, **2** *Blitum bonus-henricus*, **3** *Haloxylon persicum* and **4** *Atriplex littoralis*.

STEMS AND LEAVES

Stems apparently jointed, leaves reduced, succulent: **5** *Salicornia perennis* and **6** *Halocnemum cruciatum*. Toothed leaves: **7** *Chenopodiastrum murale*. Mealy indumentum: **8** *Chenopodium album*.

INFLORESCENCES

Densely clustered papery bracts and tepals: **9** *Amaranthus retroflexus*. Flowers in glomerules, bracts absent: **10** *Chenopodium album*. Fleshy green tepals: **11** *Suaeda aegyptiaca*.

FRUITS, FRUTING TEPALS AND BRACTS

Spine-tipped bracts, membranous tepals covering developing fruit: **12**, **13** (with mature capsule) *Salsola kali*. Fruiting bracts enclosing achenes: **14** *Atriplex prostrata* and **15** *Atriplex halimus*. Enlarged, wing-like fruiting tepals: **16** *Haloxylon salicornicum* and **17** *Noaea mucronata*.

Aizoaceae

Cornelia Klak

Stipules absent
Leaves mostly succulent
Leaves mostly opposite
Leaves simple
Ovary syncarpous

Mostly **shrubs**. **Leaves** epidermis with bladder cells. **Flowers** small to large, perianth elements (3–)5(–8); ovary (1–)5(–many) carpellate, perigynous to hypogynous or epigynous. **Fruit** usually a hygrochastic loculicidal capsule.

LEFT TO RIGHT: *Mesembryanthemum crassicaule* (Mesembryanthemoideae): note leaves succulent, flat, opposite; *Carpobrotus acinaciformis* (Ruschioideae): leaves usually trigonous, highly succulent, opposite, glaucous; *Tetragonia nigrescens*: note alternate, entire, flat leaves, often petiolate; *Lampranthus godmaniae*: note leaves opposite, trigonous, glaucous, sessile.

Characters of similar families: Kewaceae: leaves glabrous or with minute glandular hairs, alternate or subopposite, stipules persistent, flowers small, perianth simple with 2 outer perianth-segments sepaloid and the 3 inner ones petaloid. **Gisekiaceae:** flowers small, gynoecium apocarpous, fruit a cluster of mericarps. **Limeaceae:** leaves glabrous or with glandular hair, alternate to subopposite, stipules absent, flowers small, fruit as 2 woody mericarps. **Molluginaceae:** leaves glabrous, or rarely with glandular or stellate hairs or papillae, usually opposite or whorled, stipules membranous, but sometimes small, obsolete or absent, flowers small.

Shrubs, rarely herbs, trees or plants reduced to a single leaf-pair. **Leaves** usually succulent, mostly opposite, simple and entire rarely lyrate, epidermis with bladder cells, sometimes reduced, or uniform and xeromorphic, often papillate, base of petioles with stipuliform appendages or leaves sessile, sometimes with a connate leaf sheath. **Inflorescences** of principally dichasial pattern, complete or in various derived forms, mostly terminal, often seemingly axillary, often reduced to a single flower. **Flowers** actinomorphic, mostly bisexual, perigynous to hypogynous or epigynous, perianth elements (3–)5(–8), inner surface of upper portion petaloid or green, named calyx when petaloid elements are present; androecial elements 4–many, if many, the outer primordia developing often into petaloid organs (=petals), filaments usually free, rarely connate with the petals forming a tube, anthers dehiscing by longitudinal slits; ovary syncarpous, (1–)5(–many) carpellate, placentation axile, basal or parietal, ovules (1–)many per carpel, anacampylotropous or campylotropous, bitegmic, crassinucellate, rarely pendulous. **Fruit** mostly a hygrochastic loculicidal (rarely septicidal or xeromorphic) capsule, rarely schizocarpus, sometimes a hard and indehiscent 1-seeded nut, rarely a drupe, occasionally in aggregates, or a circumscissile capsule. **Seed** mostly more or less ovoid, rarely arillate.

Literature: Hartmann (1988, 2017); Herre (1971); Klak *et al.* (2007, 2017).

The family includes ca. 123 genera and ca. 1,880 species; worldwide distribution in the subtropics, with most genera and species confined to the western parts of Namibia and South Africa. Major genera (including 80–200 species) are *Antimima*, *Conophytum*, *Delosperma*, *Drosanthemum*, *Lampranthus*, *Mesembryanthemum* and *Ruschia*. Several of the larger groups need revision.

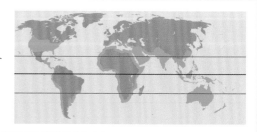

HABIT AND LEAVES

Usually perennial shrubs: **1** *Meyerophytum meyeri*. Plants compact: **2** *Cheiridopsis pillansii*. Leaves pubescent: **3** *Aizoon glinoides*. Leaves papillate: **4** *Cleretum bruynsii*.

INFLORESCENCES AND FLOWERS

Inflorescence often reduced to a single flower: **5** *Jordaaniella dubia*. Stamens and petals of staminodial origin, petals coloured: **6** *Mesembryanthemum hypertrophicum*. Filamentous staminodes collected into a central cone around the stamens: **7** *Drosanthemum asperulum*. Perianth sepaloid and green outside, and petaloid and coloured inside: **8** *Tetragonia herbacea*.

FRUIT

Side view of closed (left) and top view of open hygrochastic capsule (right): **9** *Cheiridopsis alba-oculata*. Immature fruit: **10** *Hymenogyne glabra*. Fruit winged: **11** *Anisostigma schenckii*. Fruit compound, woody, spiny: **12** *Tribulocarpus dimorphanthus*.

Cactaceae

Nigel P. Taylor

Areoles present, ± spiny
Leaves if present alternate
Stems and/or leaves fleshy
Ovary inferior, 5–∞-carpellate
Perianth segments 5–∞

Usually spiny stem **succulents**, rarely leafy. **Stems** terete, ribbed or tuberculate, bearing felted areoles, these originating branches and flowers. **Flowers** ± sunken into the receptacle (pericarpel). **Fruit** berry-like, spiny or naked.

LEFT TO RIGHT:
Echinocactus horizonthalonius;
Cylindropuntia versicolor.

Characters of similar families: Euphorbiaceae: stems lacking areoles, with milky latex, flowers apetalous, 3-carpellate, ovary superior, fruits explosively dehiscent. **Apocynaceae (Asclepiadoideae)**: stems lacking areoles, with latex, flowers 5-merous, seeds with hair tufts. **Didiereaceae**: with seasonally borne leaves, stem areoles when present not felted, flowers 4–5-merous. **Asteraceae (*Kleinia*)**: stems lacking areoles and spines, flowers in capitula, seeds (fruit) with hair tufts.

Usually spiny stem **succulents**, rarely with fleshy leaves when in growth (*Pereskia, Leuenbergeria, Quiabentia, Pereskiopsis, Austrocylindropuntia* and *Maihuenia*). **Tree-** or **shrub-like**, columnar (unbranched), globose and solitary or caespitose, semi- or quite geophytic, or epiphytic and either climbing or pendent and emitting aerial roots. **Stems** terete, triangular or flattened, smooth, ribbed or tuberculate, growth indeterminate, or determinate and segmented, bearing felted areoles (at least when juvenile), these originating spines, barbed glochids (Opuntioideae), hair-spines, branches and flowers. Some genera almost or completely spineless at maturity (Rhipsalideae, *Ariocarpus, Lophophora*, some *Astrophytum, Copiapoa* and *Eriosyce*). A **pseudo-inflorescence**, either lateral or terminal, composed of bristles and wool, is present in a few genera. **Flower** ± sunken into the receptacle (pericarpel), the latter naked, scaly, areolate-spiny, felted or with hair-spines, often tubular. Perianth undifferentiated into sepals and petals, the segments usually very numerous (10+), variously coloured or white (rarely bluish-tinged), stamens and stigma-lobes 5–∞. **Fruit** berry-like, spiny or naked, dehiscent or not, usually with fleshy funicular pulp, rarely dry. **Seeds** usually many, mostly 0.5–3 mm, sometimes arillate, in Opuntioideae the bony aril encloses the entire seed.

Literature: Hunt (2016); Hunt *et al.* (2006); Hunt & Taylor (2011).

130 genera and 1,500 species; largely in the Americas (0–4,800 m), but *Rhipsalis baccifera* in Africa, Madagascar, Seychelles and Sri Lanka. Large genera are *Opuntia* (150+ species), *Mammillaria* (150+ species), *Echinopsis* (*sensu lato*, 100+ species), *Parodia* (50+ species), *Echinocereus* (50+ species) and *Selenicereus* (40+ species). Some larger-growing genera have edible fruit with high vitamin C content: *Opuntia, Selenicereus* (*Hylocereus*) and *Cereus*.

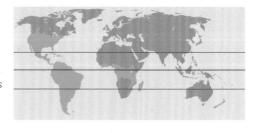

HABIT, LEAVES AND FLOWERS

All flowers sunken into the receptacle. Non-succulent climber with broad leaves: **1** *Pereskia aculeata*. Cushion plant with small evergreen leaves: **2** *Maihuenia poeppigii*. Segmented subshrub: **3** *Cumulopuntia leucophaea*. Flattened stem-segments and glochids, lacking true spines: **4** *Opuntia basilaris*. Small columnar ribbed cactus: **5** *Bergerocactus emoryi*. Dwarf ribbed cactus: **6** *Echinocereus rigidissimus*. Giant ribbed tree-cactus: **7** *Carnegiea gigantea*. Barrel-shaped cactus with waxy ribbed stems: **8** *Copiapoa cinerea*. Large barrel-shaped cactus with a very spiny ribbed stem: **9** *Ferocactus gracilis*. Dwarf tuberculate cactus: **10** *Escobaria vivipara*.

Hydrangeaceae

Isabel Larridon

Stipules absent
Leaves simple
Leaves opposite
Flowers bisexual
Ovary usually inferior

Woody plants. **Stipules** absent. **Leaves** simple, opposite. **Flowers** bisexual, sometimes marginal flowers sterile; perianth 4–12-merous; nectary usually present; stamens 8–200; pistil 1, styles 1–12. **Fruit** usually a capsule.

LEFT TO RIGHT:
Hydrangea kawakamii, showing habit and inflorescence, fruit and fruit cross-section, seed and embryo; *Philadelphus lewisii*, a species native to the west coast of North America from Canada to Central California.

Characters of similar families: Cornaceae: perianth 4(–5)-merous, stamens 4(–5), fruits drupaceous berries. **Viburnaceae:** flowers usually 5-merous, fruits drupes or berries. **Caprifoliaceae:** flowers usually 5-merous, synsepalous, synpetalous. **Rubiaceae:** interpetiolar stipules, margins always entire, flowers usually 5-merous.

Trees, **(sub)shrubs** or **vines**, evergreen or deciduous. **Stipules** absent. **Leaves** simple, usually opposite; blade sometimes palmately lobed, margins entire, serrate, serrulate, dentate, denticulate, or crenate; venation pinnate or acrodromous. **Inflorescences** terminal or axillary, cymes, panicles, racemes, or corymbs, or flowers solitary. **Flowers** generally bisexual, radially symmetric; marginal flowers sometimes sterile and bilaterally symmetric with enlarged petaloid sepals; perianth with hypanthium completely adnate to ovary or adnate to ovary proximally, free distally; sepals 4–12, distinct or connate basally; petals 4–12, usually connate basally; nectary usually present; stamens 8–200, usually distinct, sometimes connate proximally; anthers dehiscing by longitudinal slits; pistil 1, 2–12-carpellate; ovary partly or completely inferior, 1–12-locular, placentation usually axile proximally, parietal distally; ovules 1–50 per locule, anatropous; styles 1–12, distinct or connate; stigmas (1–)2–12. **Fruit** usually a capsule, dehiscence septicidal, loculicidal, interstylar, or intercostal, rarely a berry. **Seeds** 1–50 per locule.

Literature: Freeman (2016); Plants of the World Online (2021).

Hydrangeaceae has its largest diversity in North America, also native from Asia to the Pacific Islands, and from Central to South America. Introduced in Europe. The family includes 9 genera (ca. 217 species). Major genera include *Deutzia* (ca. 73 species), *Philadelphus* (ca. 52 species) and *Hydrangea* (ca. 80 species).

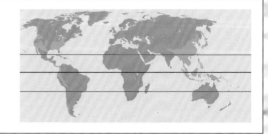

HABIT

Shrub: **1** *Hydrangea sargentiana*. Subshrub: **2** *Kirengeshoma palmata*.

LEAVES

Leaf with denticulate leaf margins: **3** *Deutzia calycosa*. Leaf lower surface: **4** *Deutzia calycosa*. Leaf blade palmately lobed: **5** *Kirengeshoma palmata*.

INFLORESCENCES

Diversity of inflorescences: **6** *Deutzia longifolia*, **7** *Hydrangea sargentiana*, **8** *Carpenteria californica* and **9** *Philadelphus lewisii*.

FLOWERS AND FRUITS

4 sepals and 4 petals: **10** *Philadelphus zhejiangensis*. Many stamens: **11** *Philadelphus insignis*. Marginal flowers with enlarged petaloid sepals: **12** *Hydrangea kawakamii*. Capsules: **13** *Deutzia crenata*. Unusually berries: **14** *Hydrangea arguta*.

Cornaceae

Isabel Larridon

Stipules absent
Leaves simple
Leaves opposite
Flowers bisexual
Ovary inferior

Woody plants. **Leaves** simple, opposite. **Flowers** usually bisexual, perianth 4(–5)-merous, stamens 4(–5), pistil 1, style 1, stigmas 2; ovary inferior. **Fruit** usually a drupaceous berry, sometimes a syncarp.

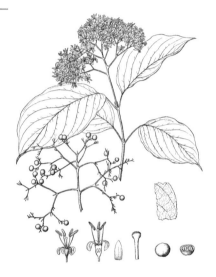

LEFT TO RIGHT:
Cornus macrophylla, showing habit, inflorescence and fruits; *Cornus wilsoniana*, collected by the famous plant hunter Ernest Henry Wilson.

Characters of similar families: Hydrangeaceae: perianth 4–12-merous, stamens 8–200, fruits usually capsules. **Garryaceae:** evergreen, dioecious. **Rubiaceae:** interpetiolar stipules. **Viburnaceae:** usually 5-merous, usually 3–5 carpels.

Trees or **shrubs**, rarely rhizomatous herbs, mostly deciduous, rarely dioecious. **Stipules** absent. **Leaves** simple, usually opposite; blade margins entire; venation pinnate or parallel; often pubescent with hairs unbranched or T-shaped; fine threads present when leaves torn in *Cornus*. **Inflorescences** terminal, rarely axillary, cymose, paniculate, corymbose, umbellate, or capitulate; bracts minute, not petaloid, early caducous, or 4(–6) and usually showy. **Flowers** usually bisexual; perianth and androecium epigynous; hypanthium completely adnate to ovary; sepals 4(–5), distinct or slightly connate; petals 4(–5), distinct, valvate, creamy white or yellow, rarely dark reddish purple or partially dark reddish purple; nectary present, intrastaminal; stamens 4(–5), distinct, free; anthers dehiscing by longitudinal slits; pistil 1, (1–)2(–4)-carpellate, ovary inferior, (1–)2(–4)-locular, placentation apical; ovules 1 per locule, apotropous to epitropous; style 1; stigmas 2. **Fruit** a drupaceous berry, white, blue, red, or black, berries distinct or fused into a fleshy syncarpous compound fruit. **Seeds** 1–2 per fruit.

Literature: Murrell & Poindexter (2016); Plants of the World Online (2021); Qiuyan & Boufford (2005).

Two genera: *Cornus* (ca. 60 species) and *Alangium* (ca. 57 species). *Cornus* is distributed in northern boreal and temperate regions (Eurasia to Indochina and North America), and also occurs at high elevations in (sub)tropical regions from South Sudan to southern Tropical Africa, and throughout Central America to Bolivia. *Alangium* occurs in the Old World (sub)tropics.

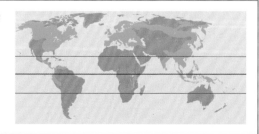

HABIT

Tree: **1** *Cornus mas*. Shrub: **2** *Cornus alternifolia*. Rhizomatous herb: **3** *Cornus unalaschkensis*. Fine threads visible in torn leaves: **4** *Cornus sanguinea*.

INFLORESCENCES AND FLOWERS

Inflorescence showing umbels of yellow flowers: **5** *Cornus officinalis*. Showy inflorescence bracts: **6** *Cornus hongkongensis*. Flowers 4-merous: **7** *Cornus officinalis* (yellow) and **8** *Cornus controversa* (white). Ovary inferior: **9** *Cornus sanguinea*.

FRUITS

White fruits: **10** *Cornus alba*. Red fruits: **11** *Cornus mas*. Black fruits: **12** *Cornus sanguinea*. Fleshy syncarpous compound fruits: **13** *Cornus kousa*.

Polemoniaceae

Clare Drinkell

Stipules absent
Flowers bisexual
Flowers actinomorphic
Perianth parts fused
Ovary superior

Herbs. **Stipules** absent. **Leaves** alternate, opposite or whorled. **Inflorescence** a cyme. **Flowers** actinomorphic, 5-merous; sepals connate; corolla tubular, stamens alternate with lobes; ovary superior, 3 fused carpels. **Fruit** a dehiscent capsule.

LEFT TO RIGHT:
Gilia achilleifolia: inflorescence a globose, densely packed cyme;
Polemonium caeruleum;
Linanthus androsaceus.

Characters of similar families: Apocynaceae: sap present, style simple, fruit a berry, drupe or dehiscing pair of follicles. **Campanulaceae**: leaves rarely compound, inflorescence typically racemose, ovary usually inferior, 2–5-locular. **Diapensiaceae**: mostly glabrous, flowers solitary. **Gentianaceae**: flowers 4–5-merous, carpels usually 2, fused to form a single locule, stigma capitate or 2-lobed, seeds irregularly angled.

Usually perennial or annual **herbs**, rarely lianas, shrubs or small trees. Indumentum hairy, glandular, viscid, rarely glabrous. **Stipules** absent. **Leaves** simple to compound, usually alternate, sometimes opposite or whorled, entire to deeply lobed, petiolate or sessile, rarely with a modified tendril (*Cobaea*). **Inflorescence** a simple to compound **cyme** in a determinate, corymbose, racemose, paniculate or capitate arrangement, usually terminal, rarely solitary (*Cobaea*). **Flowers** usually actinomorphic, bisexual, 5-merous (4–6-merous in some *Linanthus*); sepals usually connate, persistent, membranous between the lobes; corolla tubular, usually funnel-form to salverform, lobes convolute in bud; stamens 5, arranged alternately with corolla lobes, filaments fused basally to the tube at equal to unequal lengths, exserted to included; anthers basifixed to dorsifixed, dehiscing by longitudinal slits; ovary superior with 3 fused carpels (2 in *Navarretia*), each carpel with 1–many ovules; a nectariferous disk surrounding the ovary base; single style usually with 3 stigma branches. **Fruit** a dehiscent loculicidal capsule. **Seeds** 1–many.

Literature: Grant (2003); Johnson (2009); Porter & Johnson (2000); Rhui-cheng & Wilken (1995); Wilken (2004).

27 genera with ca. 400 species arranged in 3 subfamilies. Mainly in northern temperate regions, with most diversity occurring in woodlands and shrublands of western North America. The family was formerly placed in Solanales on the basis of floral characters, but is now included in Ericales, close to Fouquieriaceae, on the basis of DNA and morphology. Some genera are cultivated as garden ornamentals.

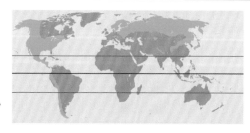

HABIT AND LEAVES

Usually herbs. Cushion-like habit growing in a dry rocky area: **1** *Phlox condensata*. Alternate leaves growing in moist ground: **2** *Polemonium reptans*. Leaves simple: **3** *Collomia grandiflora*. Leaves pinnately dissected: **4** *Gilia tricolor*.

INFLORESCENCES AND FLOWERS

Sepals membranous between lobes: **5** *Linanthus bigelovii*. Corolla lobes convolute in bud: **6** *Phlox divaricata*. Tubular corolla: **7** *Ipomopsis aggregata*. 3-lobed style: **8** *Polemonium viscosum*. Nectary disk: **9** *Ipomopsis tenuituba*.

FRUIT

Capsule: **10** *Aliciella latifolia* subsp. *latifolia*. Loculicidal dehiscence: **11** *Gilia angelensis*.

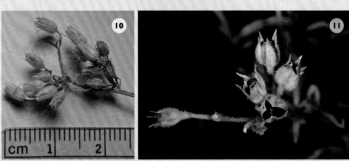

163

Primulaceae

Clare Drinkell

Stipules absent
Leaves simple
Flowers bisexual
Flowers actinomorphic
Ovary superior

Herbs, shrubs. Leaves often forming basal rosettes. Hairs often present. **Inflorescences** often scapose. **Flowers** bisexual, actinomorphic, 5-merous; stamens equal to and opposite the corolla lobes; ovary usually superior. **Fruit** a capsule.

LEFT TO RIGHT:
Lysimachia nutans;
Primula farinosa.

Characters of similar families: Campanulaceae: latex often present, flowers actino- or zygomorphic, ovary usually inferior, with annular disk. **Saxifragaceae:** stamens 5 or 10, anthers 2-locular; intrastaminal disk present, ovary inferior to superior.

Perennial or annual **herbs**, suffrutescent (woody-based), rarely tuberous, sometimes tufted or cushion-forming, **shrubs**, rarely small trees, rarely aquatic. Glandular hairs often present. **Stipules** absent. **Leaves** simple, often forming a basal rosette, alternate, opposite or whorled; margins entire to serrate or rarely finely pinnate (*Hottonia*); sometimes glandular punctate. **Inflorescences** often scapose, terminal or axillary, solitary, racemose, spicate or umbellate, bracts usually present. **Flowers** bisexual, actinomorphic, rarely zygomorphic, usually 5-merous; sepals basally connate, persistent in fruit; petals basally connate, corolla tube short to very long, campanulate to salverform, lobes usually overlapping, entire, notched, fringed or reflexed (*Cyclamen* and some *Primula*); stamens the same number as and opposite the petals, spreading or connivent forming a cone, rarely with staminodes, often heterostylous, anthers opening by slits; ovary superior, rarely semi-inferior (*Samolus*), 1-locular, placentation free central; several to many ovules; style 1, stigma not lobed, capitate. **Fruit** a dehiscing capsule, sometimes a drupe, with persistent calyx. **Seeds** few to numerous, angular.

Literature: Alpine Plant Encyclopaedia (online); Anderberg (2004); Chi-ming & Kelso (1996); Ståhl & Anderberg (2004).

55 genera with ca. 3,000 species in four subfamilies. Most temperate species are in the Primuloideae subfamily. Temperate species of Myrsinoideae tend to be herbs or shrubs, rarely small trees. Mainly occurring in northern temperate regions, preferring alpine habitats of meadows or rocky crevices, or wet, boggy habitat and woodlands. Some genera are widely cultivated and important horticulturally.

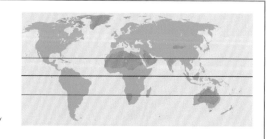

HABIT AND LEAVES

Cushion forming: **1** *Androsace chamaejasme*. Erect scapose herb: **2** *Primula pulverulenta*. Leaves opposite: **3** *Lysimachia tenella*. Rosettes of dentate leaves: **4** *Dionysia mozaffarianii*. Leaves glandular punctate: **5** *Lysimachia arvensis*.

INFLORESCENCES AND FLOWERS

Zygomorphic flowers: **6** *Coris monspeliensis*. Actinomorphic flowers. Corolla lobes fringed: **7** *Soldanella alpina*. **8** Reflexed petals, stamens meeting at the tips: *Cyclamen purpurascens*. Stamens spreading: **9** *Lysimachia clethroides*.

FRUIT

Globose capsule with coiling peduncle: **10** *Cyclamen hederifolium*. Fruit a dehiscent capsule with a single bract on the pedicle: **11** *Samolus valerandi*. Fruit a globose drupe: **12** *Ardisia crenata*.

Ericaceae

Timothy Utteridge

Sap absent
Stipules absent
Leaves simple
Leaves spirally arranged
Perianth parts fused

Stipules absent. **Leaves** spiral, often evergreen. **Flowers** sympetalous, corolla urceolate or tubular and often brightly coloured; stamens 5–10, with various appendages and anthers dehiscing through pores. **Fruit** with persistent calyx.

LEFT TO RIGHT:
Enkianthus deflexus;
Erica hillburttii;
Rhododendron insigne.

Characters of similar families: Plantaginaceae (e.g. *Penstemon*): anthers dehiscing by slits, fruit a septicidal capsule. **Myrtaceae:** leaves opposite with distinct intra-marginal veins and pellucid gland dots, stamens generally numerous. **Theaceae:** petals free, stamens numerous, anthers dehiscing through slits, ovary superior without persistent calyx at apex.

Herbs, subshrubs, shrubs or **trees** (rarely scandent), sometimes mycotrophic (Monotropoideae), aromatic compounds sometimes present (e.g., methyl salicylate in *Gaultheria*). **Sap** not present. **Hairs** present, usually simple, *Rhododendron* often with conspicuous stellate hairs or scales on the young parts. **Stipules** absent. **Leaves** variable but usually spiral, rarely pseudoverticillate or opposite, margins entire or toothed; plane and flat or recurved, sometimes extremely so and 'ericoid' (though an ambiguous term); venation pinnate sometimes with 3 or more secondary veins from the base (triplinerved), or closely parallel. **Inflorescences** various, axillary or terminal, often racemose but also paniculate, fascicles etc. **Flowers** bisexual (Ericoideae unisexual), pedicels often articulated, sometimes with 2 bracteoles; calyx and corolla usually 4–5-merous, corolla urceolate or tubular, often attractive and colourful; stamens usually 5–10, dehiscing with apical pores, anthers with appendages (awns); ovary superior to inferior, 5–10-locular. **Fruit** with persistent calyx, a capsule, berry or drupe. **Seeds** 1–10 to numerous.

Literature: Fang *et al.* (2005); Oliver (2012); Stevens *et al.* (2004a); Tucker (2009).

About 125 genera with at least 4,200 species; widely distributed in temperate and subarctic regions, also in the montane tropics. Several species rich groups: *Erica* – 840 species (Europe to southern Africa); *Rhododendron* – ca. 1,100 species (northern hemisphere to Asian montane tropics, >500 species endemic to China); and *Vaccinium* – 440 species (cosmopolitan). Important horticultural and food crops (e.g., blueberry and cranberry). Now includes the Epacridaceae, Monotropoideae and Pyroloideae (the latter two sometimes treated as Pyrolaceae).

HABIT

Trees, note peeling bark: **1** *Arbutus menziesii*. Simple leaves with serrate margins: **2** *Arbutus xalapensis*. Spirally arranged leaves, terminal bud with protective bracts: **3** *Rhododendron excellens*. Dense scaly indumentum on young parts, note dried bracts: **4** *Rhododendron makinoi*.

FLOWERS AND FRUIT

Tubular corollas, note 'ericoid' leaves: **5** *Erica caffra*. Urceolate corollas: **6** *Arbutus unedo* and **7** *Pieris japonica*. 'Epacroids': **8** *Epacris* sp. and **9** *Dracophyllum secundum*. Large showy, spreading corollas: **10** *Rhododendron scabrum* and **11** *Rhododendron makinoi*. Berries: **12** *Arbutus unedo* and **13** *Vaccinium corymbosum*. Dehiscent capsule: **14** *Rhododendron occidentale*.

Theaceae

Timothy Utteridge

Leaves simple
Leaves alternate
Flowers actinomorphic
Perianth parts free
Ovary superior

Leaves simple often with toothed or serrate margins, stipules lacking. **Flowers** axillary, solitary; sepals and petals free; stamens numerous. **Fruit** a dehiscent capsule.

LEFT TO RIGHT:
Camellia sp.;
Schima sinensis

Characters of similar families: **Actinidiaceae:** climbers or trees, flowers lacking bracteoles, fruit a berry. **Clusiaceae:** yellow to red sap present, leaves opposite, entire. **Pentaphylacaceae:** leaves entire, filaments short and anthers long, fruit various but often berries. **Sapotaceae:** white sap present, stipules often present, margins entire. **Symplocaceae:** leaves drying yellow, ovary partly inferior, fruit a drupe.

Trees or shrubs, without sap. **Hairs** present, simple, especially conspicuous on the young parts which often have silver or gold silky hairs. **Stipules** absent. **Leaves** simple; petiolate, spiral or alternate, usually coriaceous with toothed or serrate margins. **Flowers** solitary (rarely with up to 3 flowers) and axillary; sometimes with a pair of bracteoles below the flower (and leaving a pair of scars on the pedicel); usually large and bisexual, actinomorphic with bracteoles, sepals and petals spirally arranged and intergrading; sepals 5, free, imbricate; petals 5 (or more, sometimes up to ca. 14), free, imbricate; stamens numerous, usually fused at the base and adnate to the corolla or petals, filaments long and anthers short, connective not exserted; ovary superior, usually 5-locular with 2–few axillary ovules. **Fruit** usually capsular, fleshy and leathery or dry and woody rarely drupaceous and indehiscent. **Seeds** flattened and sometimes winged.

Literature: Min & Bartholomew (2007); Prince (2007, 2009); Stevens *et al.* (2004b).

Theaceae has 11 genera and ca. 305 species; primarily in the Asian tropics and subtropical America (e.g. *Camellia* with ca. 190 species), but also in the USA and the Neotropics. The cultivated monotypical genus *Franklinia*, native to Georgia in the USA, is extinct in the wild.

HABIT AND LEAVES

Tree with smooth bark: **1** *Stewartia sinensis*. Alternate, coriaceous leaves with serrate margins, note young flower buds covered with silky hairs: **2** *Camellia* sp. and **3** *Camellia japonica*.

FLOWERS

Solitary flowers with overlapping sepals and petals and numerous stamens: **4** *Camellia saluenensis*, **5** *Franklinia alatamaha*, **6** *Stewartia pseudocamellia* and **7** *Polyspora* cf. *axillaris*. Mature flower with stamens fused at the base: **8** *Camellia pitardii*. Mature flower with numerous stamens and free styles: **9** *Stewartia sinensis*.

FRUIT

Mature oblongoid fruit before dehiscing: **10** *Polyspora* cf. *axillaris*. Immature globose fruit: **11** *Camellia japonica*. Mature dehisced fruits: **12** *Stewartia monadelpha*.

Rubiaceae

Sally Dawson

Stipules present
Leaves simple
Leaves opposite
Leaf margins entire
Ovary inferior

Stipules interpetiolar. **Leaves** opposite simple, entire. **Flowers** actinomorphic, corolla tubular; stamens epipetalous, alternipetalous, number equalling the corolla lobes; ovary inferior.

LEFT TO RIGHT: *Rubia tinctorum*; *Paederia foetida*; *Coprosma arborea*; *Galium maximoviczii*.

Characters of similar families: Myrtaceae: leaves with pellucid (clear) gland dots, interpetiolar stipules lacking, flowers usually with many anthers. **Apocynaceae:** abundant white exudate present, interpetiolar stipules usually lacking, ovary superior. **Oleaceae:** interpetiolar stipules lacking, stamens 2 (rarely 4), ovary superior.

Trees, **shrubs**, woody or herbaceous **climbers**, **annual** or **perennial herbs**, rarely **epiphytes** or **aquatics**, sometimes with thorns or hooks, stems square or terete. Raphides sometimes present. **Stipules** interpetiolar rarely intrapetiolar, undivided, bifid or multi-lobed, or stipules leaf-like; colleters often on stipules or calyces. **Leaves** simple, never compound; opposite decussate, distichous, or whorled; sometimes anisophyllous; margins entire or rarely with hooks or lobes, sometimes domatia present, sometimes bacterial nodules present. **Inflorescences** axillary or terminal, often on reduced shoots, thyrsoid, highly variable or flowers solitary. **Flowers** bisexual, or unisexual, often heterostylous, or with secondary pollen presentation, sometimes wind pollinated; commonly 4–5-merous, actinomorphic; calyx, corolla lobes and stamens usually equal in number, calyx lobes sometimes uneven to few enlarged, stamens alternipetalous; ovary inferior, rarely superior, (1)2–(many)-locular, placentation axile, sometimes parietal, or rarely both; style simple, entire or bilobed or as a pollen presenter. **Fruit** indehiscent fleshy, or leathery or woody berries or drupes, or dehiscent capsules or mericarps, sometimes fused. Seeds variable, (1)2–many, rarely winged or with elaiosomes; endosperm present, rarely absent.

Literature: Bremer & Manen (2000); Davis *et al.* (2009); Chen *et al.* (2011); Soza & Olmstead (2010); Tutin *et al.* (1976).

A cosmopolitan family of ca. 600 genera and ca.14,000 species. The mainly herbaceous tribe Rubieae is the predominant tribe in Temperate regions. *Galium* is the largest temperate genus in terms of number of species and it also has the most widespread distribution.

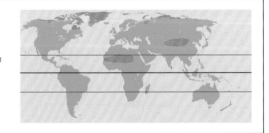

HABIT

Herbs: **1** *Cruciata laevipes*. Herbs with leaf-like stipules and angled stem: **2** *Galium aparine*. Shrubs with interpetiolar stipules and domatia in the leaves: **3** *Coprosma robusta*. Climbers and scrambling herbs and shrubs: **4** *Rubia peregrina*. Rarely trees in temperate areas: **5** *Emmenopterys henryi*.

INFLORESCENCES AND FLOWERS

Wind pollinated, female flower, colleters at site of the fallen stipule: **6** *Coprosma repens*. Secondary pollen presentation, exserted styles: **7** *Phuopsis stylosa*. Globose inflorescence: **8** *Cephalanthus occidentalis*. Bilobed style: **9** *Serissa japonica*.

FRUIT

Dry fruit with hooked bristles: **10** *Galium aparine*. Fleshy fruit: **11** *Coprosma robusta*. Purple-blue fruit with persistent calyx: **12** *Coprosma propinqua*. Maturing dry fruits on spherical heads: **13** *Cephalanthus occidentalis*.

Gentianaceae

Daniel Cahen

Stipules absent
Leaves simple
Leaves opposite
Flowers actinomorphic
Ovary superior

Herbs, glabrous. **Stipules** absent. **Leaves** simple, opposite, with entire margins. **Flowers** with a persistent tubular calyx and a showy tubular corolla; stamens adnate to the corolla tube. **Fruit** a septicidal capsule.

LEFT TO RIGHT:
Gentiana lutea;
Gentiana lawrencei;
Swertia iberica

Characters of similar families: Apocynaceae: exudate present, anthers forming a cone around a swollen stylar head or adnate to it and forming a gynostegium, fruits often paired dehiscent follicles. **Campanulaceae:** leaves alternate, ovary semi-inferior. **Rubiaceae:** stipules present, ovary inferior. **Plantaginaceae, Scrophulariaceae:** leaves sometimes alternate, commonly toothed, lobed or dissected, flowers often zygomorphic, stamens often markedly unequal. **Solanaceae:** leaves alternate, calyx often accrescent, fruit often fleshy.

Herbs (partially mycoheterotrophic in *Bartonia* and *Obolaria*), sometimes subshrubs, annual, biennial or perennial, usually glabrous. **Stipules** absent. **Leaves** simple, generally opposite and decussate, rarely alternate or whorled, generally sessile, often connate at base (upper leaves perfoliate in *Blackstonia*), sometimes in a basal rosette, margin entire, venation pinnate. **Inflorescence** cymes, thyrses and verticillasters, terminal and/or axillary, cymes sometimes arranged in heads, corymbs, umbels or racemes, or flowers solitary. **Flowers** usually actinomorphic and bisexual, 4–5(–12)-merous, sepals, petals and stamens isomerous; calyx tubular, usually persistent; corolla tubular, right-contorted in bud, larger than calyx, showy, usually marcescent, variable in shape but most often bell- to funnel-shaped, often plicate at sinuses, petal margins fringed (*Gentianopsis*), scales or fimbriae sometimes present at the base of the corolla lobes (some *Gentiana* and *Gentianella*), nectaries sometimes present on the petal surface (subtribe Swertiinae); stamens adnate to corolla, alternating with petals, sometimes inserted in the sinuses of the corolla lobes; carpels 2, connate, usually forming a single locule, placenta sometimes thickened and ovaries pseudobilocular; style 1 with a simple to bilobed stigma; ovary superior, ovules many, placentation parietal. **Fruit** usually a septicidal capsule, 2-valved. **Seeds** many, small, sometimes winged.

Literature: Adams (1996); Ho & Pringle (1995); Marais & Verdoorn (1963); Struwe & Pringle (2018); Tutin *et al.* (1972).

A family of 102 genera and about 1,750 species distributed worldwide but most diverse in temperate regions and tropical mountains. The family includes plants with pharmacologically active compounds that are extensively used in medicine. Also grown as ornamentals (*Eustoma* as cut flowers, *Exacum affine* as potted plants, *Gentiana* and others in gardens).

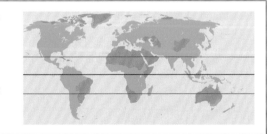

HABIT AND LEAVES

Herbs: **1** *Gentiana acaulis*, **2** *Comastoma tenellum*, **3** *Centaurium erythraea* and **4** *Frasera speciosa*. Mycoheterotroph: **5** *Obolaria virginica*. Leaves simple, opposite: **6** *Gentiana asclepiadea* and **7** *Blackstonia perfoliata*.

FLOWERS AND FRUIT

Corolla plicate at sinuses: **8** *Gentiana verna*. Petals fringed: **9** *Gentianopsis crinita*. Petal base fimbriate: **10** *Gentianella campestris*. Nectaries on petals: **11** *Swertia perennis*. Capsules: **12** *Centaurium erythraea*.

Apocynaceae

Nina Davies

Sap or exudate present
Stipules absent
Leaves simple
Leaves opposite
Ovary superior

Flowers bisexual; **corolla** with a well-developed or short tube; **stamens** and **style head** closely associated or fused; **ovary** superior.

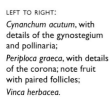

LEFT TO RIGHT:
Cynanchum acutum, with details of the gynostegium and pollinaria;
Periploca graeca, with details of the corona; note fruit with paired follicles;
Vinca herbacea.

Characters of similar families: Rubiaceae: no sap or exudate, leaves with interpetiolar stipules, ovary inferior. **Gentianaceae:** no sap or exudate, leaves sometimes 3- or 5-nerved at base, fruit a capsule, rarely a berry. **Myrtaceae:** leaves with pellucid gland dots, flowers usually with many anthers, ovary inferior. **Oleaceae:** no sap or exudate, leaves often sub-opposite, flowers usually 4-merous, stamens 2. **Convolvulaceae:** leaves alternate, corolla with mid-petaline bands, fruit usually a capsule.

Climbers or **lianas**, **perennial herbs**, **trees** or **shrubs**, more rarely subshrubs or succulents. Sap or exudate almost always present, usually white. **Stipules** absent (mostly). **Leaves** simple, opposite usually decussate or whorled (rarely alternate); lamina margins entire. **Inflorescence** cymose, racemose or flowers solitary. **Flowers** actinomorphic, bisexual; sepals 5, free; corolla tubular, with 5 lobes, usually convolute in bud; stamens 5, inserted on the corolla tube and alternating with the corolla lobes; or corolla mostly valvate in bud, stamens 5, fused to the expanded stylar head to form a compound structure, the gynostegium; corona lobes often present on gynostegium, occasionally on corolla, or on both; pollen grouped into waxy pollinia, linked by corpusculum secreted from stylar head and distributed in pairs (pollinaria); disk present or absent; ovary superior (rarely semi-inferior), 2(–8) fused carpels, or 2 free carpels united by the style; 1–2 locular with 1–many ovules in each locule; style simple with a swollen apex. **Fruit** usually 1–2 dehiscent follicles but also a berry, drupe or capsule. **Seeds** with endosperm; mostly with a coma of hairs at one or both ends; coma absent from berries and drupes.

Literature: Endress *et al.* (2018); Li *et al.* (1995); Plants of the World Online (2019); Stevens (2001 onwards).

A large cosmopolitan family of 366 genera and ca. 6,450 species, most diverse in the tropics. Apocynaceae are often poisonous with some genera used for their medicinal properties and as poisons. Species of *Amsonia*, *Asclepias*, *Mandevilla*, *Nerium*, *Plumeria*, *Stephanotis* and *Vinca* are used as ornamentals.

HABIT AND LEAVES

Climber: **1** *Trachelospermum asiaticum*. Perennial herb: **2** *Vincetoxicum hirundinaria*. Shrub: **3** *Nerium oleander*. Succulent: **4** *Stapelia similis*. Leaves simple, opposite, entire: **5** *Vinca major*. White sap or exudate present: **6** *V. major*.

FLOWERS

Corolla tubular, with 5 lobes, usually convolute in bud: **7** *Trachelospermum jasminoides* and **8** *Amsonia tabernaemontana*. Fringed corona inserted at mouth of corolla tube: **9** *Nerium oleander*. Corolla mostly valvate in bud, corona present on gynostegium: **10** *Vincetoxicum hirundinaria*. Corona inserted at mouth of the corolla tube: **11** *Periploca laevigata*.

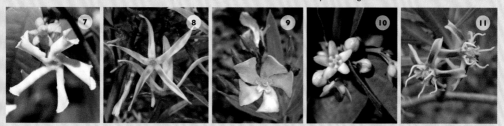

FRUIT

Single follicle: **12** *Asclepias speciosa*. Paired fruits: **13** *Vinca major* 'oxyloba' and **14** *Periploca laevigata*. Follicular dehiscence: **15** *P. laevigata*. Berry: **16** *Acokanthera oppositifolia*.

Boraginaceae

Gemma Bramley

Stipules absent
Leaves simple
Leaves alternate
Flowers actinomorphic
Ovary superior

Leaves simple, alternate, often rough hairy. **Inflorescences** cymose, often scorpioid. **Flowers** actinomorphic, 5-merous, sympetalous; style gynobasic; ovary superior. **Fruit** a schizocarp, (1–2)4 nutlets.

LEFT TO RIGHT: *Anchusa granatensis*; *Echium pininana*; *Anchusa granatensis*.

Characters of similar families: Heliotropiaceae: style terminal, fruit nutlets or drupe. Hydrophyllaceae: style terminal, fruit a capsule. Lamiaceae: leaves opposite, usually zygomorphic corolla. Solanaceae: fruit many-seeded.

Herbs rarely shrubs; stipules absent. **Leaves** alternate, simple; lamina often rough to hispid, hairs often with swollen bases, margins entire. **Inflorescence** terminal or axillary, usually scorpioid, monochasial or dichasial, occasionally solitary, more often in lax to condensed cymes. **Flowers** blue, white, yellow or pink, often changing colour as they mature, usually actinomorphic, bisexual, usually 5-merous; calyx lobes almost entirely fused to united at base only; corolla sympetalous, 5-merous, tubular, campanulate, salver-shaped, funnel-form or rotate; stamens typically 5, attached to the corolla tube; anthers exserted or not; ovary superior with 2 carpels, 4-locular; style gynobasic, stigma bilobed to capitate. **Fruit** (1–2)4 nutlets, smooth, winged or often ornamented with glochidiate spines. **Seed** with or without endosperm.

Literature: Boraginales Working Group (2016); Valentine & Chater (1972).

A family of 94 genera and ca.1,800 species; Boraginaceae is presented here *sensu stricto* (several large genera are now excluded — *Heliotropium* (Heliotropiaceae), *Phacelia* (Hydrophyllaceae)). The family is predominantly northern temperate but extends to mountainous areas of the tropics. Major genera are *Cynoglossum*, *Cryptantha* and *Myosotis*.

HABIT AND LEAVES
Usually herbs: **1** *Amsinckia lycopsoides* and **2** *Echium virescens*. Leaves with rough hairs: **3** *Pentaglottis sempervirens*.

FLOWERS
Inflorescence often scorpioid, monochasial (**4** *Trigonotis cavaleriei*) or dichasial (**5** *Symphytum orientale*), occasionally solitary (**6** *Lithodora prostrata*). Flowers actinomorphic. **7** *Anchusa aegyptiaca*. Rarely slightly zygomorphic: **8** *Echium vulgare*. Flowers often changing colour as they mature: **9** *Pulmonaria officinalis*.

NUTLETS
Often ornamented: **10** *Cynoglossum officinale*, **11** *Adelinia grande* and **12** *Bothriospermum chinense*.

Convolvulaceae

Ana Rita Simões

Stipules absent
Leaves alternate
Flowers bisexual
Flowers actinomorphic
Ovary superior

Herbs, **shrubs**, **twiners**. **Leaves** alternate. **Flowers** actinomorphic, sympetalous, 5-merous; sepals usually free and overlapping; corolla funnel-shaped to tubular with mid-petaline bands; ovary superior, syncarpous; anthers not connivent, longitudinally or spirally dehiscing. **Fruit** often 3–4-valved dry capsule.

  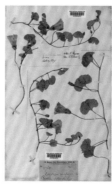 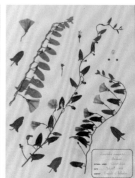

LEFT TO RIGHT:
Dichondra repens;
Convolvulus arvensis;
Calystegia soldanella;
Convolvulus arvensis.

Characters of similar families: Cucurbitaceae: tendrils present, ovary inferior. **Solanaceae:** sepals fused, anthers connivent, occasionally dehiscing by pores, staminodes present; fruit many-seeded, often a berry. **Apocynaceae:** leaves opposite; corolla tubular, mid-petaline bands absent, gynoecium apocarpous, fruit a follicle, drupe or berry.

Herbs, **shrubs**, **frequently twining or prostrate**, less often erect, rarely trees or leafless parasites. **Sap or latex** milky, often present. **Leaves** simple, weakly to deeply lobed, or compound, alternate; margins usually entire. **Inflorescences** usually axillary cymes, sometimes capitate, or solitary, often bracteate. **Flowers** bisexual, actinomorphic, sympetalous, 5-merous; sepals usually free and overlapping, persistent, sometimes enlarged in fruit; corolla campanulate to funnel-shaped, tubular or salverform; entire to shallowly or deeply lobed, corolla lobes bearing mid-petaline bands, often with distinct colour, texture or indumentum from the remainder of the corolla; stamens 5, usually included; anthers free (not connivent), longitudinally or spirally dehiscing (never by pores); ovary superior, 2(3–5)-carpellate, 2–4-lobed, styles 1 or 2, terminal or gynobasic; stigmas biglobular, elliptic or filliform. **Fruit** typically a (3–)4-valved capsule, more rarely a single-seeded utricle or fleshy berry, or nut-like. **Seeds** 1–4, sometimes hair-bearing.

Literature: Heywood *et al.* (2007); Simões & Staples (2017); Stefanović *et al.* (2003); van Ooststroom (1953); Wood *et al.* (2015).

61 genera and ca. 1,900 species widespread in tropical and temperate regions, but significantly more diverse in the tropics. The main genera in temperate regions are *Calystegia*, *Convolvulus*, *Cuscuta*, *Dichondra* and *Ipomoea*.

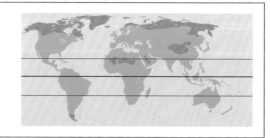

HABIT AND LEAVES

Prostrate: **1** *Convolvulus arvensis* and **2** *Dichondra sericea*. Climbing or twining: **3** *Ipomoea cairica* and **4** *D. sericea*. Leaves simple: **5** *Calystegia sepium* and **6** *Dichondra macrocalyx*. Leaves compound: **7** *I. cairica*.

FLOWERS

Infundibuliform: **8** *Convolvulus hermanniae* and **9** *Calystegia sepium*. Tubular: **10** *Ipomoea cordatotriloba*. Midpetaline bands (white): **11** *Calystegia soldanella*. Anthers spiralling: **12** *Decalobanthus peltatus*.

FRUIT

Dehiscent capsule 4-valved (**13** *Ipomoea kituiensis* and **14** *Distimake cissoides*) or 3-valved: (**15** *Ipomoea rubens*).

Solanaceae

Timothy Utteridge

Leaves simple
Leaves alternate
Flowers bisexual
Corolla actinomorphic
Ovary superior

Habit various including **herbs**, **shrubs** and **small trees**. **Stipules** absent. **Flowers** tubular, calyx and corolla usually 5-lobed; stamens usually 5; ovary 2–5-locular. **Fruit** a fleshy berry or capsule. **Seeds** numerous.

LEFT TO RIGHT: *Scopolia carniolica*; *Solanum jamesii*; *Solanum americanum*.

Characters of similar families: Boraginaceae: inflorescences cymose, often 1-sided, fruit (1–2)–4 nutlets. **Convolvulaceae:** stems twining, corolla with mid-petaline bands, seeds 1–4 often hairy. **Campanulaceae:** latex often present, ovary usually inferior.

Herbs, **climbers**, **shrubs** and **small trees**, often with prickles. **Sap** absent. **Hairs** usually present, diverse including simple, stellate and dendroid. **Stipules** absent. **Leaves** simple, occasionally compound, alternate, spirally arranged or subopposite, petiolate or sessile, entire or lobed, sometimes modified to spines, domatia rarely present. **Inflorescences** terminal, axillary or leaf-opposed, racemose, paniculate, umbelliform or flowers solitary. **Flowers** usually bisexual, actinomorphic or zygomorphic; calyx tubular or campanulate, variously fused with (3–)5(–9) lobes, persistent in fruit; corolla variously fused, campanulate, funnel-shaped, urn-shaped or salver-shaped; stamens usually 5, inserted inside or at throat of corolla tube, alternating with lobes, equal or unequal in length, filaments usually free, anthers bilocular or unilocular, dehiscing by slits or pores; ovary superior, 2–5-locular, ovules usually numerous, placentation axile, style simple, stigma capitate. **Fruit** usually a 2-locular berry or capsule. **Seeds** usually numerous, often flattened.

Literature: D'Arcy (1991); PBI *Solanum* Project (2021); Purdie *et al.* (1982); Särkinen *et al.* (2013); Zhang *et al.* (1994).

A family of 100 genera and between 3,000 and 4,000 species. Worldwide but most diverse in the Americas; the largest genus is *Solanum* with ca. 1,200 species. Globally economically important as foods, including potatoes (*Solanum tuberosum*), tomatoes (*Solanum lycopersicum*) and chillies (*Capsicum* spp.), tobacco, ornamentals and medicines, and as model organisms for genetic research.

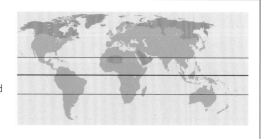

HABIT AND LEAVES

Often herbs and shrubs: **1** *Nicandra physalodes*, **2** *Lycium shawii* and **3** *Fabiana imbricata*. Leaves alternate, simple: **4** *Solanum dulcamara* and **5** *Atropa bella-donna*.

FLOWERS

Actinomorphic: **6** *Solanum elaeagnifolium*, **7** *Physochlaina orientalis* and **8** *Lycium barbarum*. Zygomorphic: **9** *Hyoscyamus niger*. Calyx persistent in fruit: **10** *Atropa bella-donna*.

FRUIT

Berries: **11** *Solanum stelligerum*, **12** *Lycium ferocissimum*, **13** *Nicandra physalodes* (one persistent calyx lobe removed) and **14** *Solanum elaeagnifolium*.

Oleaceae

Gemma Bramley

Stipules absent
Leaves simple or compound
Leaves opposite
Flowers actinomorphic
Ovary superior

Stipules absent. **Leaves** opposite. **Flowers** actinomorphic, bi- or unisexual; **corolla** usually 4-merous; stamens 2 (rarely 4); ovary superior, 2 carpels, 2 locules. **Fruit** a 1-seeded drupe, berry or samara, rarely a capsule.

LEFT TO RIGHT: *Fraxinus ornus*; *Ligustrum obtusifolium* subsp. *obtusifolium*; *Syringa afghanica*.

Characters of similar families: Rubiaceae: interpetiolar stipules, stamens usually equal in number to corolla lobes, ovary inferior. Apocynaceae: white sap, twisted buds, stamens 5. Lamiaceae: flowers often zygomorphic; stamens 4 (2); fruit usually 4 nutlets.

Trees, **shrubs** or **woody climbers**; branches often lenticellate. **Stipules** absent. **Leaves** simple, trifoliolate or imparipinnate; opposite, sometimes subopposite, rarely alternate. **Inflorescences** in cymes, panicle, racemes, umbels or fascicles; terminal or axillary. **Flowers** bisexual, rarely unisexual (plants monoecious, dioecious or polygamodioecious), actinomorphic; calyx 4(–16)-lobed, rarely absent; corolla 4-lobed (to 16-lobed in *Jasminum*), usually sympetalous, sometimes tube extremely short with lobes joined in pairs at base of staminal filament, rarely absent (some *Fraxinus*); stamens 2 (rarely 4), attached to corolla tube when sympetalous; ovary superior, syncarpous with 2 carpels and 2 locules; style terminal; stigma 2-lobed or capitate. **Fruit** usually a 1-seeded berry, drupe or samara, a loculicidal capsule or a circumscissile capsule (*Menodora*). Seeds 1(–2).

Literature: Green (2004); Chang *et al.* (1996); Wallander (2014); Wallander & Albert (2000).

An almost cosmopolitan family with 24 genera and ca. 615 species. *Fraxinus* species are used for timber; *Jasminum*, *Syringa*, *Ligustrum*, *Osmanthus*, *Chionanthus* and *Forsythia* are common ornamentals. The olive, *Olea europaea*, is cultivated in Mediterranean climates.

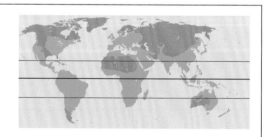

HABIT

Trees and shrubs: **1** *Fraxinus excelsior* and **2** *Osmanthus delavayi*. Climbers: **3** *Jasminum nudiflorum*. Twigs lenticellate: **4** *Picconia excelsa*. Leaves opposite: simple (**5** *Syringa vulgaris*), trifoliolate (**6** *Jasminum azoricum*) or pinnate (**7** *F. excelsior*).

FLOWERS

Actinomorphic: 4-merous (**8** *Chionanthus virginicus* and **9** *Cartrema americana*) or rarely >4-merous (**10** *Jasminum azoricum*). Perianth absent: **11** *Fraxinus excelsior*. Stamens 2: **12** *Forsythia viridissima* and **13** *Ligustrum vulgare*.

FRUITS

Drupe: **14** *Olea europaea* and **15** *Phillyrea latifolia*. Samara: **16** *Fraxinus ornus*. Capsule: **17** *Syringa pubescens*.

Plantaginaceae

Gemma Bramley

Stipules absent
Leaves usually simple
Flowers bisexual
Perianth parts fused
Ovary superior

Leaves alternate and spiral or opposite; usually simple. **Flowers** bisexual, zygomorphic, 2-lipped, occasionally actinomorphic; stamens (2)–4; ovary superior, 2 fused carpels, ovules numerous, placentation axile. **Fruit** a capsule.

LEFT TO RIGHT:
Penstemon sp.;
Veronica x *divergens*;
Veronica orientalis subsp.
carduchorum.

Characters of similar families: Scrophulariaceae s.s.: primarily southern hemisphere; (sub)actinomorphy more common. **Lamiaceae:** up to 4 nutlets, or a drupe with 1–4 seeds. **Orobanchaceae:** hemi- or holoparasites.

Herbs, **sometimes shrubs**; terrestrial, rarely aquatic; autotrophic; hairs simple, often glandular, with gland head not vertically divided; stipules absent. **Leaves** alternate and spiral, opposite or basally opposite, apically alternate, usually simple, cauline or basal; margins entire to subentire, toothed or lobed. **Inflorescence** variable; sometimes flowers solitary. **Flowers** bisexual, occasionally monoecious or dioecious; zygomorphic, less often actinomorphic, rarely reduced; calyx lobes fused, 4–5-merous; corolla usually 5–(8)-merous, can appear 4-merous due to the fusion of the upper lobes, often 2-lipped, tubular to various degrees, sometimes with basal nectar spur or personate; stamens (2)4, didynamous to equal, adnate to corolla, 0–1(–3) staminode(s) present; anthers 2-locular, sacs divergent, opening by 2 longitudinal slits, or a single inverted slit if apical portion of anther sacs adnate; ovary superior, 2-carpellate, carpels fused, placentation axile, ovules (1)–numerous, anatropous or hemitropous; stigma bilobed or capitate. **Fruit** a capsule, usually septicidal or loculicidal, sometimes poricidal or circumsessile. **Seeds** (1)–numerous, winged or angular.

Literature: Albach *et al.* (2005); Fischer (2004); Freeman *et al.* (2019b).

At least 90 genera and 1,900 species; mostly in the northern hemisphere. Plantaginaceae expanded significantly following the disintegration of the large and polyphyletic Scrophulariaceae. *Veronica, Russelia, Angelonia* and *Penstemon* are popular ornamentals; *Plantago* is unusual with ±parallel venation and wind-pollinated flowers; *Callitriche* are aquatics; *Antirrhinum* is a model organism for floral development; and *Digitalis* species are used medicinally. Several species are noxious weeds.

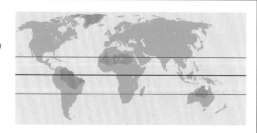

HABIT AND LEAVES

Usually terrestrial herbs: erect (**1** *Penstemon fruticosus* and **2** *Digitalis lanata* or prostrate (**3** *Cymbalaria muralis*). Aquatic: **4** *Callitriche* sp. Shrubs: **5** *Veronica odora*. Leaves alternate, cauline: 6 *Linaria purpurea*. Leaves opposite: **7** *Gratiola officinalis*. Basal rosette: **8** *Plantago lanceolata*.

FLOWERS

Corolla zygomorphic: tubular (**9** *Digitalis purpurea* and **10** *Penstemon* sp.), with spur (**11** *Linaria vulgaris*) or personate (**12** *Antirrhinum latifolium*). Actinomorphic: **13** *Erinus alpinus*. Reduced perianth parts: **14** *Plantago lanceolata*. Stamens 4: **15** *Penstemon unilateralis*. Stamens 2: **16** *Veronica chamaedrys*.

FRUIT

Fruit a capsule, style persistent: **17** *Linaria purpurea*, **18** *Veronica agrestis* and **19** *Veronica salicifolia*. Seeds tiny, numerous: **20** *Digitalis purpurea*.

Scrophulariaceae

Gemma Bramley

Stipules usually absent
Leaves simple
Flowers bisexual
Perianth parts fused
Ovary superior

Leaves simple, alternate or opposite. **Flowers** bisexual, actinomorphic to zygomorphic, regular to 2-lipped; stamens 2, 4 or 5; ovary superior, 2 fused carpels, ovules numerous, placentation axile. **Fruit** usually a capsule.

LEFT TO RIGHT:
Nemesia bodkinii;
Verbascum coromandelianum;
Diascia austromontana.

Characters of similar families: Lamiaceae: up to 4-nutlets, or a drupe with 1–4 seeds. **Orobanchaceae:** hemi- or holoparasites. **Plantaginaceae:** tend to be more diverse in the northern Hemisphere, actinomorphy less common. **Boraginaceae:** leaves rough hairy, flowers actinomorphic, 5-merous. **Onagraceae:** stamens usually 8, ovary inferior.

Herbs, sometimes **shrubs**; terrestrial, rarely sub-aquatic (*Limosella*); autotrophic; hairs usually simple, if glandular the gland head flattened, discoidal, with vertical partitions; stipules absent (interpetiolar lines present in *Buddleja*). **Leaves** alternate or opposite, or basally opposite, apically alternate, usually simple, cauline or basal; margins entire to subentire, toothed or lobed; lamina can be gland-dotted (especially in Myoporeae). **Inflorescences** racemes or thyrses, terminal flower usually absent. **Flowers** bisexual, commonly (sub)actinomorphic, or zygomorphic (particularly in *Eremophila* and *Scrophularia*); calyx lobes fused, 3–5-merous; corolla usually 4–5-merous, 2-lipped or regular; stamens 2, 4 or 5 (–8 in *Myoporum*), didynamous to equal, adnate to corolla, 0 or 1(–3) staminode(s) present; anthers 2-locular, sacs confluent, opening by distal slit, base not sagittate; ovary superior, 2-carpellate, carpels fused, placentation axile or apical (free-central in *Limosella*), ovules (1)–numerous, anatropous or hemitropous; style bilobed or capitate. **Fruit** a loculicidal or septicidal capsule, a berry, or drupe-like. **Seeds** (1)–numerous.

Literature: Fischer (2004); Oxelman *et al.* (2005); Rabeler *et al.* (2019); Tank *et al.* (2006); Thiesen & Fischer (2004).

Scrophulariaceae is a family of 59 genera and ca. 1,880 species. Scrophulariaceae s.s. includes Buddlejaceae and Myoporaceae, but many former genera are now recognised as Plantaginaceae and Orobanchaceae. Most diversity is in the southern Hemisphere: four out of eight tribes are almost entirely South African, one is Australian (Myoporeae), and only Scrophulariaeae (including *Scrophularia* and *Verbascum*) is predominantly northern temperate.

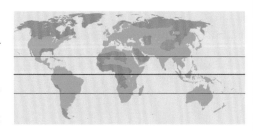

HABIT AND LEAVES

Erect or creeping herbs, leaves opposite or rosette-forming: **1** *Verbascum thapsus*, **2** *Anticharis glandulosa* and **3** *Scrophularia* sp.
Shrubs: **4** *Buddleja davidii* and **5** *Myoporum laetum*

FLOWERS

Actinomorphic: **6** *Zaluzianskya ovata* and **7** *Buddleja davidii*. Zygomorphic: **8** *Nemesia fruticans* and **9** *Scrophularia nodosa*.
Subactinomorphic: stamens 5 (**10** *Verbascum phlomoides*) or 4 (**11** *Selago densiflora*).

FRUIT

Immature capsule: **12** *Scrophularia nodosa*. Mature capsule: **13** *Verbascum levanticum*. Drupe-like: **14** *Eremophila longifolia*.

Lamiaceae

Gemma Bramley

Stipules absent
Leaves simple or compound
Leaves opposite
Flowers bisexual
Ovary superior

Leaves opposite. **Inflorescence** cymose or thyrsoid; calyx persistent, often accrescent; corolla tubular, usually zygomorphic; stamens (2)4, exserted. **Fruit** schizocarp with (1)–4 nutlets, sometimes a drupe.

LEFT TO RIGHT: *Melissa officinalis*; *Stachys serbica*; *Marrubium incanum*; *Nepeta grandiflora*.

Characters of similar families: **Rubiaceae**: stipules present, leaf margins entire, ovary inferior. **Boraginaceae**: leaves alternate, flowers usually actinomorphic. **Solanaceae**: leaves alternate, fruit many-seeded. **Scrophulariaceae, Plantaginaceae, Orobanchaceae**: fruits usually many-seeded capsules.

Herbs, **shrubs**, **trees** or **lianas**; herbaceous stems usually quadrangular; often the crushed foliage aromatic or foetid; stipules absent. **Leaves** simple, more rarely compound, opposite (decussate); lamina margins often toothed or lobed; hairs often present, simple to complex. **Inflorescences** usually 1–many flowered cymes of various forms, particularly verticillasters or thyrses; terminal or axillary, usually with bracts. **Flowers** usually zygomorphic, more rarely actinomorphic; calyx 2–5-lobed, rarely unlobed or with more than 5 lobes, campanulate, or funnel-shaped, or tubular, persistent, often accrescent; corolla more or less 5-lobed, usually bilabiate unless actinomorphic; stamens 4, or 2 by abortion and staminodes usually present; ovary superior, unlobed to deeply 4-lobed, 2-carpellate, often appearing 4-locular by intrusions of the ovary wall constituting 'false septa'; 2 ovules per carpel (1 in each 'locule'), lateral attachment, placentation axile; style gynobasic or terminal, not persistent (except in Australia). **Fruit** more or less a schizocarp splitting into typically 4 nutlets, or a drupe (1)–4-seeded, both enclosed in the persistent calyx. **Seeds** endospermic to non-endospermic.

Literature: Harley *et al.* (2004); Zhao *et al.* (2021).

A family of 236 genera and ca. 7,300 species; almost cosmopolitan. Many species are well-known for their culinary and/or medicinal uses (e.g. *Lavandula*, *Thymus*, *Salvia* and *Mentha*); numerous ornamentals.

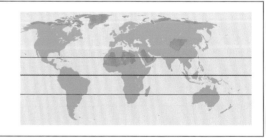

STEM AND LEAVES

Stems quadrangular: **1** *Lamium album*. Leaves opposite, decussate; usually simple, often toothed or lobed: **2** *Stachys sylvatica* and **3** *Leonurus cardiaca*. Eglandular hairs, glandular hairs or sessile glands often present: **4** *Glechoma hederacea* and **5** *Salvia pratensis*.

INFLORESCENCE AND FLOWERS

Cymes of various forms: 1-flowered (**6** *Scutellaria galericulata*), verticillasters (**7** *Lamium galeobdolon*) or thyrses (**8** *Collinsonia canadensis* and **9** *Clinopodium vulgare*). Corolla zygomorphic: 2-lipped (**10** *Prostanthera rotundifolia* and **11** *Monarda punctata*) or 1-lipped (**12** *Teucrium fruticans*). Corolla ±actinomorphic: **13** *Mentha aquatica*.

FRUIT

Calyx persistent, accrescent: **14** *Clinopodium gracile*. Schizocarp, 4-nutlets: immature (**15** *Salvia pratensis*) or mature (**16** *Stachys sylvatica*). Drupe: **17** *Callicarpa americana*.

Orobanchaceae

Iain Darbyshire and Daniel Cahen

Stipules absent
Leaves simple
Flowers bisexual
Perianth parts fused
Ovary superior

Hemi- or **holo-parasitic**, usually **herbs**; sometimes drying ink-black. **Leaves** often toothed to deeply incised. **Flowers** usually held in spikes or racemes, tubular, often 2-lipped; stamens usually 4. **Fruit** capsular, many-seeded.

LEFT TO RIGHT:

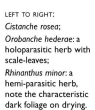

Cistanche rosea;

Orobanche hederae: a holoparasitic herb with scale-leaves;

Rhinanthus minor: a hemi-parasitic herb, note the characteristic dark foliage on drying.

Characters of similar families: Plantaginaceae and Scrophulariaceae: very similar morphologically and difficult to separate, but non-parasitic, usually not drying ink-black. Lamiaceae: inflorescence units cymose, fruit a schizocarp of up to 4 nutlets or a drupe with 1–4 seeds. Solanaceae: flowers commonly actinomorphic; stamens usually 5; fruit commonly a berry but can be capsular. Campanulaceae: ovary usually inferior or semi-inferior, with a hypanthium; in temperate regions, many species have an actinomorphic corolla.

Herbs, rarely shrubs or climbers; hemi- (then with chlorophyll) or holo- (lacking chlorophyll) parasites with haustorial attachments to the roots of a host plant, sometimes host-specific, very rarely non-parasitic; sometimes turning ink-black on drying. **Stipules** absent. **Leaves** simple, sometimes reduced to scales, opposite or spirally arranged; margins commonly toothed to deeply incised, or entire, venation pinnate. **Flowers** bisexual, held in racemes or spikes or sometimes single-flowered, terminal or axillary; calyx tubular, (2-)4- or 5-lobed; corolla tubular, often 2-lipped or less frequently subregular, 5-lobed; stamens (2)4, didynamous, thecae often awned and/or hairy, staminodes 0–2; ovary superior, with (1)2(3) fused carpel(s), style 1, stigma bilobed or capitate. **Fruit** a septicidal or loculicidal capsule, style usually persistent. **Seeds** usually numerous, angular and/or with sculptured testa, endosperm present.

Literature: Christenhusz *et al.* (2017); Crawford (2015b); Fischer (2004); Freeman *et al.* (2019a); Stevens (2001 onwards).

A cosmopolitan family, with 99 genera and more than 2,000 species, particularly rich in montane and nutrient-poor habitats in northern temperate regions. Large temperate genera include *Pedicularis* (600+ species), *Euphrasia* (250+ species), *Castilleja* (ca. 200 species) and *Orobanche* (150+ species). Some species are serious crop pests; several genera are of horticultural importance.

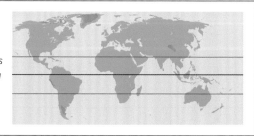

HABIT
Holoparastic herbs: **1** *Orobanche arenaria* and **2** *Cistanche phelypaea*. Hemiparasitic herbs: **3** *Pedicularis sylvatica* and **4** *Parentucellia viscosa*.

LEAVES AND INFLORESCENCES
Leaves often toothed or dissected: **5** *Pedicularis foliosa*. Flowers usually in spikes or racemes (**6** *Castilleja latifolia* and **7** *Pedicularis groenlandica*), but can appear axillary (**8** *Melampyrum pratense*).

FLOWERS
Corolla often 2-lipped: **9** *Bellardia trixago*. Stamens often appendaged and/or hairy: **10** *Euphrasia micrantha* and **11** *Rhinanthus minor*.

STIGMA AND FRUITS
Stigma bilobed (**12** *Orobanche minor*) or capitate (**13** *Lathraea clandestina*). Fruit a capsule, style often persistent (**14** *Odontites linkii*), sometimes enclosed within persistent calyx (**15** *Rhinanthus minor*).

Aquifoliaceae

Sara Edwards

Stipules present
Leaves simple
Leaves alternate
Flowers unisexual
Flowers actinomorphic

Trees or **shrubs**, dioecious. **Stipules** small. **Leaves** simple, usually alternate and leathery, margins entire to spinose. **Flowers** unisexual; actinomorphic, isomerous, inconspicuous, unisexual, ovary superior. **Fruit** a drupe, usually globose.

LEFT TO RIGHT:
Ilex paraguariensis;
Ilex mitis var. *mitis*;
Ilex chinensis.

Characters of similar families: Berberidaceae: stipules usually absent, flowers bisexual, petals with nectaries, fruits berries. **Celastraceae:** flowers bisexual, fruits berries or woody capsules. **Fagaceae** (*Quercus ilex*): flowers unisexual, monoecious, fruit a single-seeded nut in cupule. **Lauraceae:** often aromatic, stipules absent, anthers with flaps, fruit a berry in cupule. **Oleaceae** (*Osmanthus* spp.): opposite leaves, flowers usually bisexual, sepals usually fused into a tube. **Symplocaceae** (*Symplocos* spp.): stipules absent, stamens numerous, fruit often blueish.

Trees and **shrubs**, evergreen or deciduous. **Stipules** small, persistent or caducous leaving a scar. **Leaves** simple, alternate, rarely opposite or subopposite; leaf blade usually coriaceous, sometimes chartaceous or membranous, margin entire to spinose toothed. **Inflorescence** axillary, solitary, fasciculate or rarely in cymes. **Flowers** usually dioecious, actinomorphic, small, isomerous; calyx 4–5(8)-lobed, persistent, coriaceous, green; petals 4–5(8), imbricate, mostly fused at base or up to half of length, often white, cream, green, yellow, pink, or red; stamens or staminodes alternating with petals, epipetalous. **Male flowers**: anthers oblong-ovoid, introrse, longitudinally dehiscent; ovulode, subglobose or pulvinate, rostrate. **Female flowers**: staminodes sagittate or cordate; ovary superior, ovoid, 4–8(–10)-loculed, rarely pubescent; style very short to absent; stigma capitate, discoid, or columnar. **Fruit** a drupe, red, brown, black, green or yellow, globose; stigma persistent, conspicuous; exocarp chartaceous or membranous; mesocarp fleshy. **Seeds** pyrenes (1–)4–6(–23).

Literature: Chen *et al.* (2003); Galle (1997); Plants of the World Online (2019); Stevens (2001 onwards).

A family of 1 genus and ca. 500–600 species; worldwide with centres of diversity in South America and Southeast Asia. *Ilex paraguayensis* is used to make maté tea, which is commonly drunk in South America. Other species of *Ilex* are used as ornamental plants. The wood is used in carving and furniture-making as it is unusually pale.

HABIT

Tree or shrubs: **1** *Ilex aquifolium* 'Pendula' and **2** *I. aquifolium* 'Pyramidalis'. Simple alternate leaves: Shrub: **3** *Ilex aquifolium* 'Pyramidalis'.

FLOWERS

Male flowers: **4** *Ilex aquifolium*, **5** *Ilex integra* and **6** *I. aquifolium* 'Ingramii'. Female flowers: **7** *I. aquifolium* 'Scotia', **8** *Ilex pernyi* and **9** *Ilex* spp. Female flower buds with fruits: **10** *Ilex pernyi*.

FRUIT

Immature fruit: **11** *Ilex aquifolium*. Mature fruit, note conspicuous persistent stigma: **12** *Ilex pernyi* and **13, 14, 15** *I. aquifolium* (**14** 'Fructu luteo').

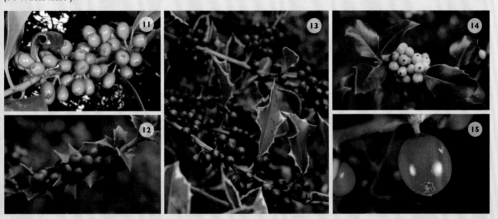

Campanulaceae

Gemma Bramley

Stipules absent
Leaves usually simple
Leaves alternate
Flowers bisexual
Ovary usually inferior

Herbs, less often shrubs, trees and lianas; milky **latex** frequent. **Stipules** absent. **Leaves** alternate, commonly simple. **Flowers** synsepalous and sympetalous, actino- or zygomorphic; ovary usually inferior, crowned with annular disk. **Fruit** a capsule or berry.

LEFT TO RIGHT: *Campanula macrostyla*; *Lobelia pedunculata*; *Lobelia bridgesii*.

Characters of similar families: Apocynaceae: leaves opposite, leaf margins entire, buds sometimes convolute, ovary superior. **Lamiaceae, Plantaginaceae, Scrophulariaceae**: leaves opposite, ovary superior. **Rubiaceae**: stipules, leaves opposite or whorled, leaf margins always entire.

Herbs, less frequently **shrubs**, **trees** or **lianas**. **Latex** often present, milky. **Stipules** absent. **Leaves** alternate, rarely opposite or whorled; simple, rarely compound (pinnate); lamina margins entire, toothed to dissected. **Inflorescences** solitary in axillary or rarely terminal position, or typically aggregated, terminal or less often axillary, commonly in cymose units but appearing racemose, paniculate or spicate. **Flowers** bisexual, actinomorphic (Campanuloideae) or zygomorphic (other subfamilies), often resupinate (Lobelioideae), mostly protandrous with specialised method of secondary pollen presentation, often blue or violet (can also be pink, red, orange, yellow, green or white); calyx synsepalous, lobes (3–)5(–10), adnate to ovary, forming a hypanthium; corolla sympetalous, lobes (4–)5(–10); stamens equalling number of corolla lobes, often attached to hypanthium or at ovary apex; filaments free or connate; ovary inferior (rarely superior), syncarpous, 2–5(–10)-locular with axile placentation (rarely 1-locular, parietal placentation), often crowned by annular disk; stigma as many lobes as locules, globose to cylindrical. **Fruit** a capsule, usually apically or laterally poricidal, or a berry. **Seeds** small, numerous.

Literature: Lammers (2007a,b).

Cosmopolitan with 84 genera and ca. 2,400 species; five subfamilies, but Campanuloideae (mostly temperate, flowers actinomorphic) and Lobelioideae (worldwide but many species neotropical, flowers zygomorphic) with ca. 90% of species. *Campanula*, *Lobelia* and *Codonopsis* are used widely in horticulture.

HABIT AND LEAVES

Usually herbs. Erect: **I** *Campanula rhomboidalis*, **2** *Campanula scouleri* and **3** *Lobelia feayana*. Creeping: **4** *Campanula poscharskyana*. Leaves simple, alternate: **5** *Campanula persicifolia*.

FLOWERS

Actinomorphic: **6** *Campanula lusitanica* and **7** *Campanula latifolia*. Zygomorphic: **8** *Lobelia laxiflora* and **9** *Lobelia erinus*. Calyx lobes forming hypanthium: **10** *Campanula rhomboidalis*. Filaments free: **11** *Platycodon grandiflorus*. Filaments connate: **12** *Lobelia tupa*. Ovary inferior: **13** *Canarina canariensis* and **14** *Campanula arvatica*.

FRUIT

Capsule: **15** *Wahlenbergia rupestris* x *violacea* and **16** *Platycodon grandiflorus*. Berry: **17** *Lobelia angulata*.

Compositae (=Asteraceae)

DJ Nicholas Hind

Stipules absent
Leaves alternate or opposite
Inflorescence compound
Corolla actinomorphic
Ovary inferior

Inflorescence a capitulum of an involucre of phyllaries enclosing florets; florets hermaphrodite or unisexual. **Corollas** actinomorphic or zygomorphic, surrounding 5 connate anthers, enclosing a style with 2 style arms. Fertile florets producing an **achene** usually with a pappus of scales or hairs.

Characters of similar families: Caprifoliaceae (Dipsacoideae): annual or perennial herbs (rarely shrubs), leaves opposite or verticillate, inflorescences of dense cymose capitula surrounded by bracts, florets all hermaphrodite, corollas usually zygomorphic, 4- or 5-lobed, stamens 5 (free), stigma simple or 2-lobed, fruit dry and indehiscent, enclosed in epicalyx, calyx persistent. **Eriocaulaceae:** annual or perennial herbs; leaves all basal, alternate, spiralled or distichous, inflorescences of capitula surrounded by bracts, florets bracteate, actinomorphic or zygomorphic, sepals 2 or 3, petals 2 or 3, stamens 2–6, ovary 2- or 3-locular, styles 1–3, fruit dry and dehiscent capsule.

Terrestrial annual, biennial or perennial **herbs**, suffrutices, **subshrubs**, or **shrubs** (rarely trees). Acaulous, or stems herbaceous or woody, simple or branched, leafy or leafless. **Stipules** absent. **Leaves** alternate, opposite or rosulate. **Inflorescences** terminal or axillary, scapose, scapiform or variously branched. Capitula homogamous and discoid or ligulate, or heterogamous and disciform or radiate, rarely radiant; involucre urceolate, campanulate, turbinate or cylindrical; phyllaries 1–many-seriate, distant or imbricate and gradate; receptacle concave, flat, convex or conical, paleaceous or epaleaceous, glabrous or pubescent. Florets small, hermaphrodite, or functionally male, female or sterile; corollas actinomorphic, filiform or tubular and (3–4) 5-lobed, or zygomorphic and bilabiate, ligulate (with 5 apical teeth), or pseudoligulate, rarely absent. Stamens 5, epipetalous; anthers usually united into cylinder, often with apical and basal appendages. Style with or without basal node, glabrous or variously pubescent, style arms 2, often well-exserted from anther cylinder. Ovary inferior, unilocular; ovary solitary, basal, anatropous. **Fruit** an achene (=cypsela), with or without an apical pappus of setae or scales, usually uni- or bi-seriate, setae capillary or flattened, smooth, finely to coarsely barbellate or plumose, or reduced to a corona or auricle.

Literature: Anderberg et al. (2007); Flora of North America Editorial Committee (2006); Funk et al. (2009); Shi et al. (2011); Tutin et al. (1976).

Senecio littoralis
(Tribe Senecioneae)

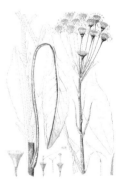

Senecio elodes
(Tribe Senecioneae)

A family of ca. 1,650 genera and ca. 25,000 species. A few genera contain apomictic microspecies that can be difficult to identify (e.g. *Taraxacum* and *Hieracium*), and could increase species numbers considerably. A global family, widespread in most temperate ecosystems, rare in aquatic environments and Antarctica. Five subfamilies, classically 14 tribes but potentially 40+ following recent phylogenetic proposals, with some tribal concepts difficult to promote.

HABIT

Herbs, short-lived perennial (**1** *Erigeron karvinskianus*), acaulous rosettiform perennial (**2** *Inula rhizocephala*), perennial (**3** *Berkheya purpurea*) or cushion-forming perennial (**4** *Cotula fallax*). Subshrub: **5** *Tanacetum densum* subsp. *amani*.

CAPITULA AND INVOLUCRES

Capitula radiate (**6** *Rudbeckia hirta*), radiant (**7** *Centaurea montana* 'Alba'), discoid (**8** *Cirsium rivulare* 'Atropurpureum') or ligulate (**9** *Helminthotheca echioides*). Synflorescence with 1-flowered capitula: **10** *Echinops ritro*. Involucres with multiseriate, gradate phyllaries (**11** *Cynara cardunculus*) or uniseriate phyllaries (**12** *Werneria orbignyana*).

FRUITING CAPITULA AND ACHENES

Achenes with conspicuous asymmetric pale-coloured basal carpopodium and awn-like pappus setae: **13** *Constancea nevinii*. Achenes with a beak beneath simple pappus, interspersed with conspicuous white receptacular paleae: **14** *Hypochaeris glabra*. Detail of beaked achenes with plumose, interwoven, pappus setae: **15** *Tragopogon pratensis*. Achenes with simple pappus setae: **16** *Hieracium tomentosum*.

Viburnaceae

Lesley Walsingham

Leaves simple or compound
Leaves opposite
Flowers bisexual
Flowers actinomorphic
Ovary inferior

Herbs, **shrubs**, or **small trees**. **Leaves** simple or compound, opposite, margins entire or serrate. **Flowers** bisexual, actinomorphic; corolla 5-merous; connate, stamens alternating with petals, ovary inferior. **Fruit** drupe or berry.

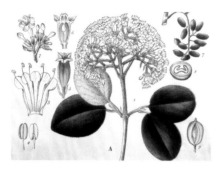

LEFT TO RIGHT:
Viburnum sp.;
Sambucus nigra;
Adoxa moschatellina.

Characters of similar families: Caprifoliaceae: flowers zygomorphic, elongate style, capitate stigma. **Rubiaceae:** interpetiolar stipules, leaf margins always entire. **Hydrangeaceae:** (*Hydrangea* similar to *Viburnum* in large outermost sterile flowers) free petals, stamens twice as many as petals. **Cornaceae:** (confused with *Viburnum*), typically 4-merous, hairs simple or 2-armed.

Herbs, perennial and rhizomatous, **shrubs**, or **small trees**. **Hairs** when present, simple or stellate. **Stipules** absent or stipule-like appendages present, often inconspicuous. **Leaves** simple or pinnately compound, rarely bipinnate, or ternate or biternately compound, opposite, sometimes whorled, or basal, usually serrate, can be lobed or entire. **Inflorescences** terminal panicles, flat-topped corymbose cymes, or glomerules, can be false umbellate, bracts and bracteoles small if present, or absent. **Flowers** bisexual, actinomorphic; sometimes outermost flowers large, zygomorphic and sterile, innermost flowers smaller, fertile (*Viburnum*); usually 5-merous, sepals 2–5, small, fused at base; corolla tubular, funneliform to campanulate, various lengths, petals 4–6, lobed, broad connate; stamens equal and alternate to petals, filaments adnate to tube, sometimes divided to base and appearing twice as many stamens to petals, anthers longitudinally dehiscent; ovary inferior to semi-inferior; 3–5 fused carpels forming one or several locules; style short, sessile or absent, stigma, 3–5-lobed. **Fruit** a drupe (*Viburnum*) or a berry, (*Sambucus*). **Seeds** (1)3(5) hard pyrenes.

Literature: Backlund & Bittrich (2016); Brummitt (2007a–c); Christenhusz *et al.* (2017); Judd *et al.* (1999); Qiner *et al.* (2011a); Walters (1976a).

Three genera and ca. 255 species, in mainly northern temperate areas, less often in the subtropics. Historically, *Viburnum* and *Sambucus* have been included in the Caprifoliaceae, or recognised as monogeneric families. Both genera are notable as ornamental shrubs, with *Sambucus* fruits traditionally used in culinary and herbal medicine practices.

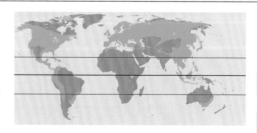

HABIT AND LEAVES

Rhizomatous herb: **1** *Adoxa moschatellina*. Small trees: **2,3** *Sambucus nigra* (leaves pinnately compound, opposite, leaflet margins serrate). Shrub: **4** *Viburnum tinus* (leaves simple, entire).

INFLORESCENCE AND FLOWERS

Sessile glomerule, filaments divided, stamens appearing twice the number of petals: **5** *Adoxa moschatellina*. Terminal flat-topped cymes: **6** *Sambucus nigra*. Stamens alternate to petals, stigma lobed: **7** *S. nigra*. Corymbose cyme, tubular flowers, inferior ovary: **8** *Viburnum* x *burkwoodii*.

FRUIT

Berries: **9** *Sambucus nigra* and **10** *Sambucus racemosa*. Drupe: **11** *Viburnum lantana* and **12** *Viburnum opulus*.

Caprifoliaceae

Lesley Walsingham

Stipules absent
Leaves opposite
Flowers bisexual
Flowers zygomorphic
Ovary inferior

Herbs, **shrubs** or **climbers**. **Stipules** absent. **Leaves** simple or pinnately compound, opposite. **Flowers** 5-merous, synsepalous and sympetalous, usually zygomorphic, tubular or 2-lipped, ovary inferior. **Fruit** a capsule, berry, drupe or achene.

LEFT TO RIGHT:
Lonicera etrusca;
Valeriana officinalis;
Dipsacus fullonum;
Lonicera etrusca.

Characters of similar families: Viburnaceae: actinomorphic flowers, lobed stigma on sessile or short style, lacking a nectary in the corolla. **Rubiaceae:** interpetiolar stipules, leaf margins always entire. **Hydrangeaceae:** petals free, stamens twice as many as petals. **Cornaceae:** flowers typically 4-merous. **Compositae:** (confused with Dipsacoideae) anthers united into cylinder; fruit an achene. **Apiaceae:** (confused with Dipsacoideae) leaves alternate, fruit a schizocarp of 2 mericarps.

Herbs, **shrubs**, and woody **climbers**. **Stipules** absent, sometimes interpetiolar stipule-like structures apparent. **Leaves** simple or pinnately compound, opposite, decussate, sometimes whorled, connate or in a basal rosette; margins entire, toothed or pinnatifid. **Inflorescences** usually axillary, thyrsoid cymes 1-, 2- or 3-flowered, bracteate, or a compact terminal capitulum (Dipsacoideae and Scabiosoideae). **Flowers** bisexual, usually zygomorphic, sometimes appearing nearer actinomorphic: campanulate to funneliform, often restricted into a narrow tube; calyx persistent, of 2–5 fused sepals, sometimes accrescent, sometimes reduced to short or long teeth only, or setae, involucel or epicalyx of connate bracteoles sometimes present; corolla sympetalous, lobes 4–5, imbricate, bilabiate unless regular, sometimes gibbous with basal nectaries, dense glandular hairs in throat, or with short abaxial spur; stamens 4–5, alternating with petals, equal in length or didynamous, filaments attached to corolla tube, anthers free, 2-locular, often prominently exserted; ovary inferior, 2–8 fused carpels, some aborting, with 1–5 fertile locules and 1–several ovules per locule, style elongate, exserted, stigma capitate. **Fruit** dry dehiscent or indehiscent capsule, berry, pitted drupe, achene or cypsela. Capsules and berries many-seeded, drupes with 2–4 pyrenes, achenes and cypselae single-seeded.

Literature: Brummitt (2007d–g); Christenhusz *et al.* (2017); Deyuan *et al.* (2011a,b); Hofmann & Bittrich (2016a,b); Judd *et al.* (1999); Mayer (2016); Moore (1976); Qiner *et al.* (2011b); Walters (1976a,b); Weberling & Bittrich (2016).

About 28 genera and ca. 825 species, predominantly temperate, mostly in the Northern hemisphere. An expanded family concept including Dipsacaceae and Valerianaceae is largely accepted. Shrubby genera *Sambucus* and *Viburnum*, which were often included in past classifications of Caprifoliaceae, are now strongly supported as Viburnaceae. *Lonicera* (honeysuckle), is known for its fragrant garden ornamentals.

HABIT AND LEAVES
Herbs: **1** *Valeriana officinalis* and **2** *Scabiosa columbaria*. Climber, leaves simple: **3** *Lonicera japonica*. Leaves pinnatiform: **4** *V. officinalis*.

INFLORESCENCES AND FLOWERS
Thrysoid inflorescences: corolla bilabiate (**5** *Lonicera etrusca* and **6** *Lonicera periclymenum*); corolla, narrow, tubular (**7** *Valeriana rubra*). Terminal involucrate capitula: **8** *Lomelosia caucasica* and **9** *Dipsacus fullonum*.

FRUIT
Dehiscent capsule: **10** *Diervilla sessilifolia*. Drupe: **11** *Symphoricarpos albus*. Achene with epicalyx of connate bracteoles: **12** *Scabiosa columbaria*. Achene and seed: **13** *Succisa pratensis*. Berries: **14** *Lonicera maackii* and **15** *Lonicera tatarica*.

Araliaceae

Alison Moore

Stipules at petiole base
Leaves alternate
Flowers bisexual
Flowers actinomorphic
Ovary inferior

Stipules or **stipule scars** conspicuous. **Leaves** often compound; alternate, spiral, crowded towards ends of branches. **Inflorescence** often with ultimate branch umbellate. **Flowers** usually small, bisexual, 5-merous; ovary inferior. **Fruits** usually drupaceous.

LEFT TO RIGHT: *Heptapleurum heterophyllum*: note compound leaves and inflorescence; *Meryta sinclairii*: note simple leaves, compound inflorescence; *Fatsia japonica*; *Gamblea ciliata*.

Characters of similar families: **Apiaceae**: stipules absent, not woody, fruit usually a 2-carpellate schizocarp. **Cornaceae**: stipules absent, leaves usually simple, opposite, inflorescences not umbellate. **Rosaceae**, *Prunus*: glands on the leaves and/or petiole, ovary superior. **Smilacaceae**: monocotyledon, paired tendrils, flowers 3-merous, ovary superior. **Pittosporaceae**: stipules absent, stamens episepalous, ovary superior.

Trees, **shrubs** or **climbers**, sometimes epiphytic, rarely **herbs**, sometimes with spines, latex absent. **Stipules** often present, intrapetiolar, ligulate or sheathing base of petiole, leaving conspicuous scars. **Leaves** pinnately to palmately compound or simple and lobed, rarely peltate, alternate (rarely opposite but not in the temperate taxa); petiolate, often clasping stem at base; margins entire to toothed; hairs often stellate, sometimes unbranched, rarely glabrous; often heteroblastic. **Inflorescences** terminal, rarely axillary, branched complexes; usually umbellate or capitate, sometimes racemose or spicate; ultimate units umbels or heads, sometimes racemes or spikes, flowers rarely solitary. **Flowers** usually bisexual (dioecious when unisexual), actinomorphic; with or without caducous bracts; pedicels often articulated; calyx absent or present only as teeth or a cup 4–5(–10); petals 4–10(–12), free or connate at base, valvate or imbricate; stamens free, isomerous (rarely more numerous) and alternipetalous in a single whorl (rarely more whorls); inflexed in bud; often inserted on a disk; ovary inferior (superior only in tropics), syncarpous with 2–5(–12) locules, 1 ovule per locule; styles free or fused into a column. **Fruit** usually drupaceous with a single-seeded pyrene, sometimes baccate, rarely a schizocarp.

Literature: Heywood (2007a); Lowry & Plunkett (2020); Mitchell *et al.* (1997); Perkins (2020); Plunkett *et al.* (2018).

Medium-sized family of 44 genera and 1,450 species. Cosmopolitan across a broad range of habitats, especially well represented in montane environments (largely tropical). The largest temperate genus is *Hydrocotyle* (175 species). Various species of *Panax* are economically and culturally important for medicinal uses (ginseng). *Aralia*, *Hedera* and *Fatsia* are commonly cultivated ornamentals.

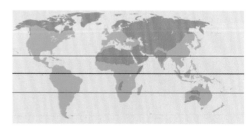

LEAVES

Simple, peltate: **1** *Hydrocotyle* sp. Simple, palmately lobed: **2** *Fatsia japonica*. Pinnately compound: **3** *Aralia elata* var. *ryukyuensis*. Stellate indumentum: **4** *Hedera helix*. Sheathing leaf bases and spines: **5** *Aralia spinosa*.

INFLORESCENCES AND FLOWERS

Umbellate globose head: **6** *Hedera helix* and **7** *Eleutherococcus sessiliflorus*. Racemose spike with capitate heads: **8** *Oplopanax horridus*. Compound umbel: **9** *Neopanax arboreus*. Persistent bracts: **10** *Aralia spinosa*. Alternipetalous stamens: **11** *Pseudopanax crassifolius*.

INFRUCTESCENCES AND FRUITS

Immature fruits, connate styles: **12** *Fatsia japonica*. Aggregated drupes on spikes: **13** *Cussonia sphaerocephala*. Fleshy drupes, persistent stigmas: **14** *Oplopanax horridus*. Panicle of umbels: **15** *Aralia cordata*. Drupe: **16** *Schefflera digitata*.

203

Apiaceae

Laura Pearce

Stipules absent
Leaves compound
Leaves alternate
Flowers bisexual
Ovary inferior

Predominantly aromatic **herbs**. **Leaves** often compound or deeply lobed and with sheathing petioles, alternate. **Inflorescence** umbellate; perianth and androecium 5-merous; petals clawed; ovary inferior. **Fruit** a schizocarp formed of 2 mericarps.

LEFT TO RIGHT:

Anethum graveolens: showing inferior ovary, stylopodium, 5-merous perianth and ridged seeds with oil ducts;

Various Apiaceae: c, carpels; cp, carpophore; d, disk (stylopodium); o, oil tubes; ov, ovary; s, stigmas;

Anthriscus sylvestris.

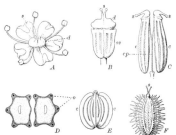

Characters of similar families: Asteraceae: stamens epipetalous, anthers often united into cylinder, ovary unilocular, fruit an achene sometimes with persistent pappus (formed of modified sepals). **Araliaceae:** woody, stipules present, ovary 2–5-carpellate, fruit a berry or a drupe. **Brassicaceae:** inflorescence corymbose or paniculate, calyx well-developed, perianth 4-merous, ovary superior, fruit a silique. **Caprifoliaceae (Dipsacoideae and Valerianoideae only):** leaves opposite, calyces modified (persistent in fruit), corolla sympetalous, fruit single-seeded.

Predominantly aromatic herbs (rarely suffrutescent or woody subshrubs, shrubs or trees). Plants glabrous, pubescent or glandular-pubescent. **Stems** erect, ascending, decumbent, prostrate or rarely creeping, often hollow at internodes. **Stipules** usually absent. **Leaves** ternately or pinnately compound, ternately, pinnately or palmately lobed, or simple, usually alternate, margins entire, toothed or deeply divided; petioles usually sheathing at base, leaves rarely sessile or perfoliate. **Inflorescences** compound-umbellate (simple-umbellate, capitulate or cymose), arranged in cymose or racemose synflorescences; umbels (and umbellules) usually subtended by an involucre (involucel) of one to many entire or dissected bracts (bracteoles). **Flowers** usually bisexual, actinomorphic (or outer flowers zygomorphic); perianth and androecium usually 5-merous; calyx lobes various, often small to obscure; petals usually basally clawed, with a narrowed, inflexed apex; stamens alternipetalous, inflexed in bud; ovary inferior, syncarpous, carpels 2; stigmas on distinct (rarely connate) stylodia, usually swollen at the base to form a nectariferous disk or stylopodium. **Fruits** usually dry and schizocarpic (rarely fleshy) with oil canals or "vittae"; at maturity the schizocarp dehisces to form 2 mericarps, usually attached to a forked or entire stalk (carpophore); mericarp ribs obscure to dentate.

Literature: Calvino & Downie (2007); Downie *et al.* (2010); Kubitzki (2018); Nicolas & Plunkett (2009).

Apiaceae has ca. 466 genera and 3,820 species, which occur worldwide but are most diverse in temperate regions of the Northern Hemisphere; habitats vary widely. Close to Araliaceae: the family and subfamily boundaries remain in flux. Major genera include *Eryngium* and *Bupleurum*. The family provides many vegetables, herbs and spices but also some very toxic species.

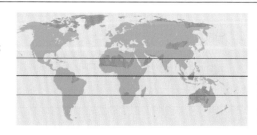

HABIT AND LEAVES

Cushion plant, leaves trilobed: **1** *Bolax gummifera*. Upright herb, leaves compound: **2** *Myrrhis odorata*. Leaves simple: **3** *Pozoa coriacea*. Sheathing petiole: **4** *Angelica sylvestris*.

INFLORESCENCES

Bracts and bracteoles: **5** *Artedia squamata*. Compound umbel: **6** *Peucedanum palustre*. Capitulum: **7** *Eryngium maritimum*. 5-merous perianth, petals clawed and inflexed: **8** *Cicuta maculata*.

INFRUCTESCENCES

Stylopodia: **9** *Bupleurum americanum*. Schizocarps: **10** *Lomatium dasycarpum*. Mature schizocarps splitting to form mericarps, and carpophores: **11** *Bupleurum fruticosum* and **12** *Heracleum maximum*.

Acknowledgements

The editors would like to thank all the contributors for their dedication to this book during a difficult period. Work commenced shortly before the full implementation of restrictions during the COVID-19 pandemic, and quickly developed into a 'lockdown' project, with the majority of the contributors working from home, juggling other responsibilities and coping with limited resources! Our completed book serves as a positive outcome from an extraordinary time in our lives; it is an example of what is possible to achieve within a group of plant enthusiasts despite challenging working conditions.

Image credits

The editors and contributors are extremely grateful to the individuals listed below for allowing use of their images. Image credits are given under the plant family and numbers in which they appear. Any images that have not been credited below are in the public domain or their copyright is owned by the Royal Botanic Gardens, Kew.

Equisetaceae: Igor Sheremetyev (1, 2, 7), Owen Clarkin (3), Lilith Ohlson (4), Riis Skov (5), Patrick Acock (6), Aurelie Grall (8), Leslie S (9). **Cyatheaceae:** Museum of New Zealand (Herb, 3, 5, 6, 8, 9), John Game (7), Adrian Tejedor (10, 11). **Dennstaedtiaceae:** Michael Sundue (1, 4, 5, 7, 8, 9), Richard Wilford (2, 6), A.V. Gilman (3). **Pteridaceae:** M. S. Beddome (drawing), Patrick Acock (1, 3), Bing Liu (2, 6, 7), John-Paul Flavell (4), Heather Brent (5), Andrew Leonard (8). **Aspleniaceae:** Richard Wilford (1-10). **Polypodiaceae:** Richard Wilford (1-11). **Cycadales:** Richard Wilford (1-6, 8, 9), Andrew McRobb (7). **Pinaceae:** Andrew McRobb (1), Kyle Port (2), Tony Kirkham (3, 8), William Friedman (4-6, 7, 9, 11), Danny Schissler (10). **Araucariaceae:** Aljos Farjon (drawing, 1, 4, 5, 7-9), Martin Xanthos (3, 6, 10-12). **Podocarpaceae:** Philippe de Spoelberch (1, 5, 7), Richard Wilford (2, 3, 6), Harry Baldwin (4). **Cupressaceae:** Harry Baldwin (1, 2), William Friedman (3-5, 7-11), Danny Schissler (6). **Taxaceae:** William Friedman (1, 3-6, 8-10), Tony Hall (2), Reni Driskill (7). **Aristolochiaceae:** Egon Krogsgaard (1, 9), Peter Gasson (2), Paul Little (3, 5), Andrew McRobb (4, 7), Henry Oakeley (6, 8), Sven Landrein (10, 12), B. Wursten (11, 13). **Magnoliaceae:** Kyle Port (1), William Friedman (2, 3, 5-9), Danny Schissler (4). **Lauraceae:** Richard Wilford (1-9, 12), Henry Oakeley (10), Rafaël Govaerts (11). **Araceae:** Ross Bayton (2, 10), Igor Sheremetyev (3), Thomas Heller (4, 5, 6, 8, 12, 13), Gwilym P. Lewis (7), Deni Bown (9), Andrew McRobb (11). **Melanthiaceae:** Ross Bayton (1, 3, 4, 6, 7), Bing Liu (2), Egon Krogsgaard (5). **Alstroemeriaceae:** (drawing) RBG Kew, Matilda Smith, R.C.H Shepherd (1, 2, 6, 11, 12) (Photo source: Australian aplant Image Index (APII), ANBG), Egon Krogsgaard (4), M. Fagg (5), Ross Bayton (7, 8), Rafael Govaerts (9). **Colchicaceae:** Rafaël Govaerts (1), Stavros Apostolou (2), Bing Liu (3, 7), Himesh Dilruwan Jayasinghe (4), Ross Bayton (5, 8, 9), Mohammed S Ali-Shtayeh, Rana M Jamous, Salam Y Abuzaitoun; Biodivers. and Environ. Res. Cen. (BERC) (6). **Smilacaceae:** Timothy Utteridge (1), Thomas Heller (2, 4), Rafaël Govaerts (3), Egon Krogsgaard (5), Ross Bayton (6), Paul Wilkin (7). **Liliaceae:** Bing Liu (3), Thomas Heller (4, 7), Ross Bayton (5), Igor Sheremetyev (6), Aleksandr Popov (8), Jared Dodson (9). **Orchidaceae:** Andre Schuiteman (1 – 11). **Iridaceae:** Daniel Cahen (2), Thomas Heller (3), Ross Bayton (4, 5, 7, 8, 9). **Asphodelaceae:** Bing Liu (1), Thomas Heller (2, 3, 5), Ross Bayton (4, 6, 8, 9, 10), Igor Sheremetyev (7), Anna Haigh (11). **Amaryllidaceae:** Ross Bayton (1, 4, 5, 6), Egon Krogsgaard (11), Daniel Cahen (12). **Asparagaceae:** (drawing) RBG Kew, Ann Farrer, Rafael Govaerts (2, 7), Egon Krogsgaard (3, 10), Ross Bayton (5, 6), Daniel Cahen (11), Ori Fragman-Sapir (12). **Palmae:** (image) F. Simonetti, W.J. Baker (1, 2, 5, 7, 8), J. Dransfield (3, 6 9), S. Andrews (4). **Juncaceae:** Igor Sheremetyev (1, 2, 5), Ori Fragman-Sapir (3, 9), Bing Liu (4, 6), Doug Goldman (7, 8). **Cyperaceae:** David Simpson (1, 2, 3, 6, 7, 8, 9, 10), Egon Krogsgaard (4), William Milliken (5). **Restionaceae:** (drawing) Walter Hood Fitch (Curtis's), Laura Jennings (1, 3, 5, 9), Egon Krogsgaard (4), John Tann (6, 8), Jean and Fred Hort (7). **Poaceae:** (drawing) Mary Maitland (Mrs

George Govan) collection, Igor Sheremetyev (1, 4), Bat Vorontsova (2, 3, 6, 11), Andrew McRobb (5, 7), Ori Fragman-Sapir (8, 10), Egon Krogsgaard (9). **Papaveraceae**: Sven Landrein (1, 11), Egon Krogsgaard (2, 3, 7, 12), Sheila Gregory (4, 5), Ori Fragman-Sapir (6), Oliver Whaley (8), Igor Sheremetyev (9), Rafaël Govaerts (10).
Berberidaceae: Ori Fragman-Sapir (1, 4, 10, 12-14), Henry Oakeley (2), Ross Bayton (3, 6), Sven Landrein (5), Andrew McRobb (7, 11), Ines Stuart-Davidson (8), Egon Krogsgaard (9). **Ranunculaceae**: Clare Drinkell (1), Sven Landrein (2, 3, 7), Andrew McRobb (4, 8, 14, 15), Henry Oakeley (5), Gemma Bramley (6), Ori Fragman-Sapir (9, 10, 12), Paul Little (11, 13). **Proteaceae**: Richard Wilford (1-10). **Buxaceae**: Harry Baldwin (1, 4, 6-9), Richard Wilford (2, 3, 5). **Hamamelidaceae**: William Friedman (1, 4-6, 8-10), Richard Wilford (2), Tony Kirkham (3), Harry Baldwin (7). **Saxifragaceae**: Richard Wilford (1-12). **Crassulaceae:** Richard Wilford (1-6, 8, 9), Rafaël Govaerts (7).
Vitaceae: Richard Wilford (1), Daniel Cahen (3), Michael Simpson (4, 7, 8), Bing Liu (5), Luis Orlando (9).
Leguminosae: Papilionoideae: (drawing) Margaret Tebbs, Daniel Cahen (1, 3, 5, 7, 12, 14, 17, 18), Ruth Clark (2, 6, 13, 14, 15, 19), Paul Wilkin (4, 11), Gwilym P. Lewis (8, 9, 10). **Polygalaceae**: Aaron Carlson (1), Laura Gaudette (2), Daniel Cahen (3, 7), Russell Best (4, 5, 12), Patrick Hacker (6), Paulmathi Vinod (8, 15), Mark Groenveldt (9), Denis Bastianelli (10), Christian Berg (11), Evan M. Raskin (13), Markus Krieger (14). **Rosaceae**: Sven Landrein (drawings, 1-14). **Rhamnaceae**: Daniel Cahen: (1-4, 7, 8, 11, 12), Jennifer Ogle (5), Leon Perrie (6), Ken-ichi Ueda (9), Matthew Fainman (10), Alexandros Quartarone (13), Don Loarie (14). **Cannabaceae**: Daniel Cahen (1, 7, 8, 11), Marina Skotnikova (2, 5, 10), Bing Lui (3), Rafaël Govaerts (4), Gennady Okatov (6), Dinesh Valke (9), Guglielmo Vacirca (12), Tatyana Malchinskaya (13), Michael Kesl (14), Vladimir Kolbintsev (15). **Moraceae**: Egon Krogsgaard (1), Daniel Cahen (2-4), Natalia Pankova (5), Lawrence Jensen (6, 9), Igor Sheremetyev (7, 11), Tatyana Malchinskaya (8, 16, 17), I. MacDonald (10), Andrey Kovalchuk (12), Andrew McRobb (13, 14), Ross Bayton (15), Rafaël Govaerts (18). **Urticaceae**: Lawrence Jensen (1, 2, 11), Leon Perrie (3), Daniel Cahen (4, 10, 12), Thomas Heller (5), Tatyana Malchinskaya (6, 9, 15), Egon Krogsgaard (7, 8), Ross Bayton (13), Hannelore Morales (14). **Fagaceae**: (drawing) RBG Kew, G. Atkinson, Miyabe – Kudo, Ross Bayton (1, 8, 9), Daniel Cahen (2, 3, 4, 5, 10, 17), Svetana Bulycheva (6), Bing Liu (7, 15), Sven Landrein (11), Alison Moore (12), Aleksei Titov (13), Andrew McRobb (14, 16). **Juglandaceae**: Andrew McRobb (1), George Hounsome (2, 6, 7, 13), Daniel Cahen (3), Saba Rokni (4, 5, 9, 14), Egon Krogsgaard (8, 11), Timothy Utteridge (10, 12). **Betulaceae**: Saba Rokni (1, 2, 7, 8, 13), Rafaël Govaerts (3, 14), George Hounsome (4, 6, 10), Daniel Cahen (5, 12), Willian Milliken (15). **Cucurbitaceae**: Egon Krogsgaard (1, 3, 5), Igor Sheremetyev (4), Himesh Dilruwan Jayasinghe (6, 10, 11), Gemma Bramley (7), Timothy Utteridge (8), **Celastraceae**: Ori Fragman-Sapir (1, 3), Timothy Utteridge (2, 4, 5, 6, 8, 11), Andrew McRobb (7), Wolfgang Stuppy (9, 10). **Oxalidaceae**: (drawing) RBG Kew, Matilda Smith, Rafaël Govaerts (1, 6, 7), Ori Fragman-Sapir (8), Daniel Cahen (10), Igor Sheremetyev (11), Valery Prokhozhy (12). **Violaceae:** RBG Kew A. Botzells (drawing), © Kim Blaxland (1, 8), Peter M. Dziuk, Minnesota Wildflowers website, https://protect-eu.mimecast.com/s/nMH2CXrRQtXR8IYf6MrEG (2, 12, 13), John and Anita Watson (3), Andrew Lane Gibson, Ohio Field Botanist (4, 5), © Phil Garnock-Jones (6, 10, 11, 15), Susan K. Wiser (7), Richard and Teresa Ware (14). **Salicaceae:** RBG Kew, A. Botzells, Nees, black and white line drawing of spherical leaf-tooth gland © Susanna Stuart-Smith (drawing), © 2014 Jan Thomas Johansson (1, 4, 5, 6, 8, 12), Peter M. Dziuk, Minnesota Wildflowers website, https://protect-eu.mimecast.com/s/nMH2CXrRQtXR8IYf6MrEG (2), SAplants (3), Sally Petitt, Cambridge University Botanic Garden (7), Used with permission from Hodel & Henrich (2020) (9, 11), Jim Conrad https://www.backyardnature.net/ (10), Kate Rabuck. Sonoma Botanical Garden, California. sonomabg.org (13) . **Euphorbiaceae s.str.**: (drawing) A. Bortzells, Daniel Cahen (1, 2), Geoff Carle, Victoria Botanic Gardens (3), Harry Rose (4), Nathan Taylor (5), Takashi Hoshide (6), Bill Higham (7), Ellura Sanctuary (8), Patrick Alexander (9), Andrea Cocucci (10), Tina Negus (11), Hans Braxmeier (12). **Linaceae**: Bing Liu (1), Igor Sheremetyev (2, 12), Hajotthu (3), Dinesh Valke (4), Ana Júlia Pereira, flora-on.pt (5), Andrea Moro, © Department of Life Sciences, University of Trieste (6, 7), Björn S (8), Egon Krogsgaard (9), Daniel Cahen (11, 15), JMK (12), Jean Weber Inra (13), Elly Vaes, RBG Kew (16). **Geraniaceae**: Nina Davies (1-3, 10, 11), Daniel Cahen (4, 5), Sheila Gregory (6, 8, 9), Saba Rokni (7), Roger & Alison Heath (12), C.

Aedo (13-15). **Lythraceae**: Richard Wilford (1, 2), Daniel Cahen (3,5-7), Egon Krogsgaard (4), Ori Fragman-Sapir (8). **Onagraceae**: Thomas Heller (1, 2), William Milliken (3), Andrew McRobb (4), Daniel Cahen (5, 7, 8), Egon Krogsgaard (6), Gemma Bramley (9). **Myrtaceae**: Ross Bayton (1, 4, 6, 7), Eve Lucas (2, 3), I. MacDonald (5), Richard Wilford (8), Félix Forest (9, 10). **Anacardiaceae**: Daniel Cahen (1, 4, 9), Ross Bayton (2, 3, 5, 6, 10), Jo Osborne (7), Sheila Gregory (8), Xander van der Burgt (11). **Sapindaceae**: Richard Wilford (1, 2, 5), Lawrence Jensen (3, 9, 10, 13), Ross Bayton (4, 11, 14), Daniel Cahen (6), Egon Krogsgaard (7), Marie Briggs (8), William Milliken (12). **Rutaceae**: Daniel Cahen (1, 6, 9), Lawrence Jensen (2, 4, 7, 14, 15), Marie Briggs (3, 10, 11), Richard Wilford (5), Ross Bayton (8, 12), Adam F. Arseniuk (13). **Malvaceae**: William Milliken (1, 2), Jim Holden (3), David Goyder (4), Andrew McRobb (5, 9), Oliver Whaley (6, 13), S. Gregory (7, 10-12), D. Zappi (8). **Thymelaeaceae**: Peter Billinghurst (1), Reuben Lim (2, 11), Alexey P. Seregin (3), Laura Jennings (4, 6), Tatiana Gerus (5), Sheila Gregory (7), Egon Krogsgaard (8), Richard Wilford (9), Ori Fragman-Sapir (10), Cerlin Ng (12). **Cistaceae**: Tony Hall (1-3, 5, 6, 10), Richard Wilford (4, 7-9). **Brassicaceae**: Igor Sheremetyev (1, 4), Shelley Heiss-Dunlop (2), Ross Bayton (3), Ori Fragman-Sapir (5, 7, 8, 11), Peter Heenan (6, 9, 15), Svetlana Kourova (10), Stephen Mifsud (12-14). **Plumbaginaceae**: William Milliken (1), Rafaël Govaerts (2), Paul Little (4), Sue Frisby (6, 9), Andrew McRobb (7), Sheila Gregory (8), Richard Wilford (10, 11). **Polygonaceae**: Egon Krogsgaard (1, 12-15, 18), Sheila Gregory (2, 5, 17), Rafaël Govaerts (3, 7), Nicky Biggs (4, 10), Henry Oakeley (6, 9, 11), Bing Liu (8), Wolfgang Stuppy (16). **Droseraceae**: Thomas Heller (1, 8), Ross Bayton (2, 5), Daniel Cahen (3, 7), Egon Krogsgaard (6). **Caryophyllaceae**: George Hounsome (1, 4, 13, 16), Sven Landrein (2, 3, 7, 18, 19), Daniel Cahen (5, 9, 10, 11, 14, 15), Jo Osborne (6), Egon Krogsgaard (12), Igor Sheremetyev (17). **Amaranthaceae**: George Hounsome (1, 9), Daniel Cahen (2, 4, 7, 8, 10, 12, 14, 15, 17), Jo Osborne (3, 6, 11, 16), William Milliken (5), Sven Landrein (13). **Aizoaceae**: (left page) Bolus Herbarium, University of Cape Town, Bolus Herbarium, University of Cape Town, Mary Maud Page, Mary Maud Page, Peter V. Bruyns (1 – 12). **Cactaceae**: Nigel Taylor (1), G. Charles (2), A. Gdaniec (3-10). **Hydrangeaceae**: Egon Krogsgaard (1), Richard Wilford (2, 5), Christophe Crock (3, 4, 6, 9-11), Y. De Smet (7), Ross Bayton (8), E. Cires Rodriguez (12), Gemma Bramley (13), G.D. Carr (14). **Cornaceae**: Richard Wilford (1, 11), Egon Krogsgaard (2, 10), Ross Bayton (3, 8), Gemma Bramley (4, 9), Christophe Crock (5-7, 13), Daniel Cahen (12). **Polemoniaceae**: John Weiser (1, 4, 8), Ross Bayton (2, 3, 6, 7), Jean Pawek (5, 10), Sven Landrein (9), Leigh Johnson (11). **Primulaceae**: Thomas Freeth (1), Ross Bayton (2), Thomas Heller (3), Richard Wilford (4, 6), Daniel Cahen (5, 7), Sven Landrein (8), Egon Krogsgaard (9), Timothy Utteridge (10), Patrick Hacker (11), Rafaël Govaerts (12). **Ericaceae**: Tony Kirkham (1, 6, 7), Richard Wilford (2, 3, 5, 12), Timothy Utteridge (4, 8-11, 13), Ross Bayton (14). **Theaceae**: Tony Kirkham (1, 5, 6, 9), Timothy Utteridge (2, 4, 8, 12), Richard Wilford (3, 11), Rogier de Kok (7, 10). **Rubiaceae**: Daniel Cahen (1), Egon Krogsgaard (2, 7, 9, 10), Lawrence Jensen (3, 6, 12), Rafaël Govaerts (4, 11, 13), Sven Landrein (8). **Gentianaceae**: Lorin Timaeus (1), Jim Morefield (2), Barry Walter (3), Luke Padon (4), E. Christina Butler (5), Ross Bayton (6), Daniel Cahen (7,8), Mark Eanes (9), Guglielmo Vacirca (10), Connie Taylor (11), Andreas Rockstein (12). **Apocynaceae**: Nina Davies (1, 3, 5, 6, 9, 13), Daniel Cahen (2, 10, 11, 14, 15), Paul Rees (4), David Goyder (7), Egon Krogsgaard (8), Andrey Zharkikh (12), Warren McCleland (16). **Boraginaceae**: Ross Bayton (1, 5, 12), Sven Landrein (2), Egon Krogsgaard (3), Gemma Bramley (4, 6, 7, 11), Sheila Gregory (8), Daniel Cahen (9), Henry Oakeley (10), Bing Liu (13). **Convolvulaceae**: Nelly Bouilhac (1, 11), Priscila Ferreira (2, 4, 5, 6, 8, 9), Ana Rita Simões (3, 12, 13), Lauren Eserman (7, 10), Fernanda Satori-Petrongari (14), Denise Sasaki (15). **Solanaceae**: Timothy Utteridge (1, 13), Thomas Heller (2), Ross Bayton (3, 6, 7, 9), Gemma Bramley (4), William Milliken (5), Egon Krogsgaard (8), Wolfgang Stuppy (10, 11, 14), Sheila Gregory (12). **Oleaceae**: Daniel Cahen (1, 3, 4, 11, 14), Ross Bayton (2, 9), Gemma Bramley (5, 13), Egon Krogsgaard (6, 8, 10), Thomas Heller (7), Richard Wilford (12, 16, 17), Rafaël Govaerts (15). **Plantaginaceae**: Richard Wilford (1, 9, 14, 16, 19), Henry Oakeley (2), Daniel Cahen (3, 11, 12, 17, 18), Ross Bayton (4, 15), Egon Krogsgaard (5, 7), Gemma Bramley (6, 8), Iain Darbyshire (10), Rafaël Govaerts (13). **Scrophulariaceae**: Sven Landrein (1), Thomas Heller (2, 5), Daniel Cahen (3, 6, 8, 9, 11, 14), Ross Bayton (4), Richard Wilford (7), Gemma Bramley (10), Sheila Gregory (12), William Milliken (13), Murray

Fagg [Australian National Botanic Gardens] (15). Lamiaceae: Gemma Bramley (1, 2, 4-7, 9, 15-17), Ross Bayton (3, 8, 11, 12), Henry Oakeley (10), William Milliken (13), Bing Liu (14). **Orobanchaceae**: Gemma Bramley (1), Iain Darbyshire (2-5, 7, 8, 13), T. Daniel [California Academy of Sciences] (6), D. Cahen (9-12, 14, 15). **Aquifoliaceae**: Rafaël Govaerts (1), Andrew McRobb (2, 3, 12-14), Thomas Heller (4), Tony Hall (5, 8), D.J.N. Hind (6, 7), Sven Landrein (9), Sara Edwards (10), Egon Krogsgaard (11, 15). **Campanulaceae**: Daniel Cahen (1, 4, 9, 10, 13), Ross Bayton (2, 3, 7, 8, 12), Gemma Bramley (5, 6), Richard Wilford (11, 16, 17), Sven Landrein (14, 15). **Compositae (=Asteraceae)**: D.J.N. Hind (1-16). **Viburnaceae**: Barry Walter (1, 5), Gemma Bramley (2, 3, 4, 8), Richard Wilford (6, 9), William Milliken (7, 12), Ross Bayton (10, 11). **Caprifoliaceae**: William Milliken (1), Richard Wilford (2, 8), Egon Krogsgaard (3, 7, 9, 11), Daniel Cahen (4), Ross Bayton (5, 14, 15), Henry Oakeley (6), Meredith Bean (10), Sven Landrein (12, 13). **Araliaceae**: Thomas Heller (1, 12), Rafaël Govaerts (2), Andrey Kovalchuk (3), Daniel Cahen (4,6), Svetlana Bulycheva (5), Bing Liu (7), Ross Bayton (8, 14, 15), Leon Perrie (9), Henry Oakeley (10), Lawrence Jensen (11, 16), Alexander Ivanov (13). **Apiaceae**: Ross Bayton (1), Esko Karjala (2), Nicholás Lavandero (3), Laura Pearce (4), Shlomit Heymann (5), Dina Nesterkova (6), Sarah Lambert (7), Bill Keim (8), Corey Raimond (9), Ron Wolf (10), Mª África de Sangenis Piñol (11), Matt Lavin (12).

References

Adams, L. G. (1996). Gentianaceae. In: A. Wilson (ed), *Flora of Australia* 28: 72–103. CSIRO, Melbourne.

Albach, D. C., Meudt, H. M. & Oxelman, B. (2005). Piecing together the "new" Plantaginaceae. *Amer. J. Bot.* 92: 297–315.

Albers, F. & Van der Walt, J. J. A. (2007). Geraniaceae. In: K. Kubitzki (ed.), *The Families and Genera of Vascular Plants, IX*: 157–165. Springer Verlag, Berlin & Heidelberg.

Allan, H. H. (1961). Brassicaceae. In: H. H. Allen, *Flora of New Zealand* 1: 174–189. Government Printer, Wellington.

Alpine Plant Encyclopaedia (online). Alpine Garden Society. www.alpinegardensociety.net/encyclopaedia/

Anderberg A. A. (2004). Primulaceae. In: K. Kubitzki (ed.), *The Families and Genera of Vascular Plants, VI*: 313–319. Springer Verlag, Berlin & Heidelberg.

Anderberg, A. A., Baldwin, B. G., Bayer, R. G., Breitwieser, J., Jeffrey, C., Dillon, M. O., Eldenäs, P., Funk, F., Garcia-Jacas, N., Hind, D. J. N., Karis, P. O., Lack, H. W., Nesom, G., Nordenstam, B., Oberprieler, C., Panero, J. L., Puttock, C., Robinson, H., Stuessy, T. F., Susanna, A., Urtubey, E., Vogt, R., Ward, J. & Watson, L. E. (2007). Compositae. In: J. W. Kadereit & C. Jeffrey, *The Families and Genera of Vascular Plants, VIII*: 61–588. Springer Verlag, Berlin & Heidelberg.

Angiosperm Phylogeny Group. (2003). An update of the Angiosperm Phylogeny Group classification for the orders and families of flowering plants: AGP II. *Bot. J. Linn. Soc.* 141 (4): 399–436.

Angiosperm Phylogeny Group. (2009). An update of the Angiosperm Phylogeny Group classification for the orders and families of flowering plants: APG III. *Bot. J. Linn. Soc.* 161 (2): 105–121.

Angiosperm Phylogeny Group. (2016). An update of the Angiosperm Phylogeny Group classification for the orders and families of flowering plants: APG IV. *Bot. J. Linn. Soc.* 181 (1): 1–20.

Appel, O. & Al-Shehbaz, I. A. (2003). Cruciferae. In: K. Kubitzki & C. Bayer (eds), *Families and Genera of Vascular Plants, V*. Springer Verlag, Berlin & Heidelberg.

Avila, N. S. (2012). Neotropical Iridaceae. In: W. Milliken, B. Klitgaard & A. Baracat, *Neotropikey—Interactive Key and Information Resources for Flowering Plants of the Neotropics*. www.kew.org/science/tropamerica/neotropikey/families/Iridaceae.htm.

Backlund, A. & Bittrich, V. (2016). Adoxaceae. In: J. W. Kadereit & V. Bittrich (eds), *The Families and Genera of Vascular Plants, XIV*: 19–29. Springer Verlag, Berlin & Heidelberg.

Baker, W. J. & Dransfield, J. (2016). Beyond *Genera Palmarum*: progress and prospects in palm systematics. *Bot. J. Linn. Soc.* 182: 207–233.

Ballard, H. E., de Paula-Souza, J. & Wahlert, G. A. (2014). Violaceae. In: K. Kubitzki (ed.), *The Families and Genera of Vascular Plants, Vol. XI*: 303–322. Springer Verlag, Berlin & Heidelberg.

Balslev, H. (1998). Juncaceae. In: K. Kubitzki (ed.), *The Families and Genera of Vascular Plants, IV*: 252–260. Springer Verlag, Berlin & Heidelberg.

Bao Bojian, Clemants, S. E. & Borsch, T. (2003). Amaranthaceae. In: Z. Y. Wu, P. H. Raven, & D. Y. Hong (eds), *Flora of China* 5: 415–429. Science Press, Beijing & Missouri Botanical Garden Press, St. Louis.

Barkworth, M. E., Capels, K. M., Long, S. & Anderton, L. K. (eds) (2006). *Magnoliophyta: Commelinidae (in part): Poaceae, part 1, Flora of North America North of Mexico* 24. Oxford University Press, New York & Oxford.

Barkworth, M. E., Capels, K. M., Long, S. & Piep, M. B. (eds) (2003). *Magnoliophyta: Commelinidae: Poaceae, Part 2. Flora of North America* 25. Oxford University Press, New York & Oxford.

Barringer, K. & Whittemore, A. T. (1997). Aristolochiaceae. In: Flora of North America Editorial Committee, *Flora of North America North of Mexico* 3. Oxford University Press, New York & Oxford.

Batdorf, L. (2005). *Boxwood Handbook*. American Boxwood Society, Boyce.

Bayer, C. & Kubitzki, K. (2002). Malvaceae. In: K. Kubitzki & C. Bayer (eds), *The Families and Genera of Vascular Plants, V*: 225–311. Springer Verlag, Berlin & Heidelberg.

Beaumont, A. J. A., Edwards, T., Manning, J., Maurin, O., Rautenbach, M., Motsi, M., Fay, M., Chase, M. & Bank, M. (2009). *Gnidia* (Thymelaeaceae) is not monophyletic: taxonomic implications for Thymelaeoideae and a partial new generic taxonomy for *Gnidia*. *Bot. J. Linn. Soc.* 160 (4): 402–417.

Bedford, D. J., Lee, A. T., Macfarlane, T. D., Henderson, R. J. F. & George, A. S. (2020). Various genera in *Flora of Australia*. Australian Biological Resources Study, Canberra. https://profiles.ala.org.au/opus/foa/

Beentje, H. (2016). *The Kew Plant Glossary*. 2nd edition. Royal Botanic Gardens, Kew.

Bittrich, V. (1993). Caryophyllaceae. In: K. Kubitzki, J. G. Rohwer & V. Bittrich (eds), *The Families and Genera of Vascular Plants*, II: 206–236. Springer Verlag, Berlin & Heidelberg.

Bittrich, V., Kubitzki, K. & Rohwer, J. G. (1993). *Families and Genera of Vascular Plants*. Springer Verlag, Berlin & Heidelberg.

Boraginales Working Group, Luebert, F., Cecchi, L., Frohlich, M. W., Gottschling, M., Guilliams, C. M., Hasenstab-Lehman, K. E., Hilger, H. H., Miller, J. S., Mittelbach, M., Nazaire, M., Nepi, M., Nocentini, D., Ober, D., Olmstead, R. G., Selvi, F., Simpson, M. G., Sutorý, K., Valdés, B., Walden, G. K. & Weigend, M. (2016). Families of the Boraginales. *Taxon* 65: 502–522.

Bown, D. (2000). *Aroids: Plants of the Arum Family*. 2nd edition. Timber Press, Portland.

Bramley, G. & Edwards, S. (2015). Aristolochiaceae. In: T. Utteridge & G. Bramley (eds), *The Kew Tropical Plant Identification Handbook*. 2nd edition. Royal Botanic Gardens, Kew.

Bredenkamp, C. L. (2000). Rhamnaceae. In: O. A. Leistner (ed.), *Seed Plants of Southern Africa: Families and Genera*. Strelitzia 10. National Botanical Institute, Pretoria. http://biodiversityadvisor.sanbi.org/wp-content/themes/bst/keys/e-Key-20160604/Families/F_Rhamnaceae.html

Breitwieser, I., Brownsey, P. J., Nelson, W. A., Smissen, R. & Wilton, A. D. (eds) (2010–2021). Meicytus. In: *Flora of New Zealand Online*. www.nzflora.info/search.html?q=melicytus.

Bremer, B. & Manen, J. F. (2000). Phylogeny and classification of the subfamily Rubioideae (Rubiaceae). *Pl. Syst. Evol.* 225: 43–72.

Brophy, J. J., Craven, L. A. & Doran, J. C. (2013). *Melaleucas: Their Botany, Essential Oils and Uses*. Australian Centre for International Agricultural Research (ACIAR), Bruce, Australian Capital Territory.

Brummitt, R. K. (2007a). Adoxaceae. In: V. H. Heywood, R. K. Brummitt, A. Culham & O. Seberg (eds), *Flowering Plant Families of the World*, pp. 26–27. Royal Botanic Gardens, Kew.

Brummitt, R. K. (2007b). Viburnaceae. In: V. H. Heywood, R. K. Brummitt, A. Culham & O. Seberg (eds), *Flowering Plant Families of the World*, pp. 331. Royal Botanic Gardens, Kew.

Brummitt, R. K. (2007c). Sambucaceae. In: V. H. Heywood, R. K. Brummitt, A. Culham & O. Seberg (eds), *Flowering Plant Families of the World*, pp. 290–291. Royal Botanic Gardens, Kew.

Brummitt, R. K. (2007d). Caprifoliaceae. In: V. H. Heywood, R. K. Brummitt, A. Culham & O. Seberg (eds), *Flowering Plant Families of the World*, pp. 86–87. Royal Botanic Gardens, Kew.

Brummitt, R. K. (2007e). Dipsacaceae. In: V. H. Heywood, R. K. Brummitt, A. Culham & O. Seberg (eds), *Flowering Plant Families of the World*, pp. 129–130. Royal Botanic Gardens, Kew.

Brummitt, R. K. (2007f). Morinaceae. In: V. H. Heywood, R. K. Brummitt, A. Culham & O. Seberg (eds), *Flowering Plant Families of the World*, pp. 219. Royal Botanic Gardens, Kew.

Brummitt, R. K. (2007g). Valerianaceae. In: V. H. Heywood, R. K. Brummitt, A. Culham & O. Seberg (eds). *Flowering Plant Families of the World*, pp. 339–340. Royal Botanic Gardens, Kew.

Brummitt, R. K. & Wilmot-Dear, C. M. (2007). Moraceae. In: V. H. Heywood, R. K. Brummitt, A. Culham & O. Seberg (eds), *Flowering Plant Families of the World*, pp. 218–219. Royal Botanic Gardens, Kew.

Calvino, C. I. & Downie, S. R. (2007). Circumscription and phylogeny of Apiaceae subfamily Saniculoideae based on chloroplast DNA sequences. *Molec. Phylogenet. Evol.* 44: 175–191.

Camus, A. (1936). *Les Chênes: Monographie du Genre Quercus*. Paul Lechevalier, Paris.

Chang, M., Qiu, L. & Green, P. S. (1996). Oleaceae. In: Z. Y. Wu & P. H. Raven, *Flora of China* 15: 272–319. Science Press, Beijing & Missouri Botanical Garden Press, St. Louis.

Chase, M., Christenhusz, M. & Fay, M. (2017). *Plants of the World: An Illustrated Encyclopedia of Vascular Plants.* University of Chicago Press, Chicago; Royal Botanic Gardens, Kew.

Chase, M. W., Zmarzty, S., Lledó, M. D., Wurdack, K. J., Swensen, S. M. & Fay, M. F. (2002). When in doubt, put it in Flacourtiaceae: a molecular phylogenetic analysis based on plastid *rbcL* DNA sequences. *Kew Bull.* 57: 141–181.

Chase, M. W., Reveal, J. L. & Fay, M. F. (2009). A subfamilial classification for the expanded asparagalean families, Asparagaceae, Amaryllidaceae and Xanthorrhoeaceae. *Bot. J. Linn. Soc.* 161: 132–136.

Cheek, M. R. (2007). Malvaceae. In: V. H. Heywood, R. K. Brummitt, A. Culham & O. Serberg (eds), *Flowering Plant Families of the World*, pp. 201–202. Royal Botanic Gardens, Kew.

Chen, J., Deng, M., Ji, Y. & Landis, J. (2020). Plastome phylogenomics of *Cephalotaxus* (Cephalotaxaceae) and allied genera. *Ann. Bot.* 127 (5): 697–708.

Chen, S., Ma, H., Feng, Y., Barriera, G. & Loizeau, P. A. (2003). Aquifoliaceae. In: Z. Y. Wu, P. H. Raven & D. Y. Hong (eds), *Flora of China* 11: 359–438. Science Press, Beijing & Missouri Botanical Garden Press, St. Louis.

Chen, S.-L., Sun, B., Liu, L., Wu, Z., Lu, S., Li, D., Wang, Z., Zhu, Z., Xia, N., Jia, L., Zhu, G., Guo, Z., Yang, G., Chen, W., Chen, X., Phillips, S. M., Stapleton, C., Soreng, R. J., Aiken, S. G., Tzvelev [Tsvelev], N. N., Peterson, P. M., Renvoize, S. A., Olonova, M. V. & Ammann, K. H. (2006). Poaceae (Gramineae). In: Z. Y. Wu, P. H. Raven & D. Y. Hong (eds), *Flora of China* 22: 1–653. Science Press, Beijing & Missouri Botanical Garden Press, St. Louis.

Chen, S., Ma, H. & Parnell, J. A. N. (2008). Polygalaceae. In: Z. Y. Wu, P. H. Raven & D. Y. Hong (eds), *Flora of China* 11: 139–159. Science Press, Beijing & Missouri Botanical Garden Press, St. Louis.

Chen, T., Zhu, H., Chen, J., Taylor, C. M., Ehrendorfer, F., Lantz, H., Funston, M. & Puff, C. (2011). Rubiaceae. In: Z. Y. Wu, P. H. Raven & D. Y. Hong (eds), *Flora of China* 19: 57–368. Science Press, Beijing & Missouri Botanical Garden Press, St. Louis.

Chen, X., Liang, S., Xu, J. & Tamura, M. N. (2000). Liliaceae. In: Z. Y. Wu & P. H. Raven (eds), *Flora of China* 24: 73–263. Science Press, Beijing & Missouri Botanical Garden Press, St. Louis.

Chen, X. Q. & Koyama, T. (2000). *Smilax*. In: Z. Y. Wu & P. H. Raven (eds), *Flora of China* 24: 96–115. Science Press, Beijing & Missouri Botanical Garden Press, St. Louis.

Chen, Y. L. & Schirarend, C. (2007). Rhamnaceae. In: Z. Y. Wu, P. H. Raven & D. Y. Hong (eds), *Flora of China* 12: 115–168. Science Press, Beijing & Missouri Botanical Garden Press, St. Louis.

Chi-ming, H. & Kelso, S. (1996). Primulaceae. In: Z. Y. Wu & P. H. Raven, *Flora of China* 15: 39–189. Science Press, Beijing & Missouri Botanical Garden Press, St. Louis.

Christenhusz, M. J., Bangiolo, L., Chase, M. W., Fay, M. F., Husby, C., Witkus, M. & Viruel, J. (2019). Phylogenetics, classification and typification of extant horsetails (*Equisetum*, Equisetaceae). *Bot. J. Linn. Soc.* 189 (4): 311–352.

Christenhusz, M. J. M. & Chase, M. W. (2014). Trends and concepts in fern classification. *Ann. Bot.* 113: 571–594.

Christenhusz, M. J. M., Fay, M. & Byng, J. W. (2018). *The Global Flora. Special Edition: GLOVAP Nomenclature Part 1* (Vol. 4). Plant Gateway Ltd, Bradford.

Christenhusz, M. J. M., Fay, M. F. & Chase, M. W. (2017). *Plants of the World, an Illustrated Encyclopedia of Vascular Plants.* Royal Botanic Gardens, Kew; University of Chicago Press.

Christenhusz, M., Reveal, J., Farjon, A., Gardner, M., Mill, R. & Chase, M. (2011). A new classification and linear sequence of extant gymnosperms. *Phytotaxa* 19: 55–70.

Clayton, W. D. & Renvoize, S. (1986). *Genera Graminum. Grasses of the World.* Her Majesty's Stationery Office, London.

Clement, W. L. & Weiblen, G. D. (2009). Morphological evolution in the mulberry family (Moraceae). *Syst. Bot.* 34: 530–552.

Crawford, F. (2015a). Amaranthaceae. In: T. Utteridge & G. Bramley (eds), *The Kew Tropical Plant Families Identification Handbook*. 2nd edition. pp. 156–157. Royal Botanic Gardens, Kew.

Crawford, F. (2015b). Orobanchaceae. In: T. Utteridge & G. Bramley (eds), *The Kew Tropical Plant Families Identification Handbook*. 2nd edition. pp. 198–199. Royal Botanic Gardens, Kew.

Culham, A. (2007a). Berberidaceae. In: V. H. Heywood, R. K. Brummitt, A. Culham & O. V. Seberg (eds), *Flowering Plant Families of the World*. pp. 59–61. Royal Botanic Gardens, Kew.

Culham, A. (2007b). Papaveraceae. In: V. H. Heywood, R. K. Brummitt, A. Culham & O. V. Seberg (eds), *Flowering Plant Families of the World*. pp. 241–243. Royal Botanic Gardens, Kew.

Culham, A. (2007c). Plumbaginaceae. In: V. H. Heywood, R. K. Brummitt, A. Culham & O. V. Seberg (eds), *Flowering Plant Families of the World*. pp. 258–259. Royal Botanic Gardens, Kew.

Culham, A. (2007d). Ranunculaceae. In: V. H. Heywood, R. K. Brummitt, A. Culham & O. V. Seberg (eds), *Flowering Plant Families of the World*. pp. 273–276. Royal Botanic Gardens, Kew.

D'Arcy, W. G. (1991). The Solanaceae since 1976, with a review of its biogeography. In: J. G. Hawkes, R. N. Lester, M. Nee & N. Estrada (eds), *Solanaceae III: Taxonomy, Chemistry, Evolution*. pp. 75–137. Royal Botanic Gardens, Kew.

Dai, L.-K., Liang, S.-Y., Zhang, S.-R., Tang, Y.-C., Koyama, T., Tucker, G. C., Simpson, D. A., Noltie, H. J., Strong, M. T., Bruhl, J. J., Wilson, K. L. & Muasya, A. M. (2010). Cyperaceae. In: Z. Y. Wu, P. H. Raven & D. Y. Hong (eds), *Flora of China* 23: 164–461. Science Press, Beijing & Missouri Botanical Garden Press, St. Louis.

Davis, A. P., Govaerts, R., Bridson, D. M., Ruhsam, M., Moat, J. & Brummitt, N. A. (2009). A global assessment of distribution, diversity, endemism, and taxonomic effort in the Rubiaceae. *Ann. Missouri Bot. Gard.* 96 (1): 68–78.

De Wilde, W. J. J. O & Duyfjes, B. E. E. (2008). Cucurbitaceae. In: T. Santisuk & K. Larsen (eds), *Flora of Thailand* 9 (4): 1–148.

Debreczy, Z. & Racz, I. (2011). *Conifers Around the World*. DebdroPress Ltd, Budapest.

DeFilipps, R. A. (1980). *Smilax*. In: T. G. Tutin, V. H. Heywood, N. A. Burges, D. H. Valentine, S. M. Walters & D. A. Webb (eds), *Flora Europaea* 5: 74. Cambridge University Press, Cambridge.

Deyuan, H., Barrie, F. R. & Bell, C. D. (2011b). Valerianaceae. In: Z. Y. Wu, P. H. Raven & D. Y. Hong (eds), *Flora of China* 19: 661–671. Science Press, Beijing & Missouri Botanical Garden Press, St. Louis.

Deyuan, H., Liming, M. & Barrie, F. R. (2011a). Dipsacaceae. In: Z. Y. Wu, P. H. Raven & D. Y. Hong (eds), *Flora of China* 19: 654–660. Science Press, Beijing & Missouri Botanical Garden Press, St. Louis.

Ding Hou (1960). Thymelaeaceae. In: *Flora Malesiana Series 1, Spermatophyta* 6 (1): 1–48.

Dorrat-Haaksma, E. & Linder, H. P. (2012). *Restios of the Fynbos*. Penguin Random House, South Africa.

Downie, S. R., Spalik, K., Katz-Downie, D. S., Reduron, J.-P. (2010). Major clades within Apiaceae subfamily Apioideae as inferred by phylogenetic analysis of nrDNA ITS sequences. *Pl. Divers. Evol.* 128: 111–136.

Dransfield, J., Uhl, N. W., Asmussen-Lange, C. B., Baker, W. J., Harley, M. M. & Lewis, C. E. (2008). *Genera Palmarum. Evolution and Classification of the Palms*. Royal Botanic Gardens, Kew.

Efloras.org. (2021). Flora of North America. [online] Available at: < http://www.efloras.org/flora_page.aspx?flora_id=1 >

Endress, M. E., Meve, U., Middleton, D. J. & Liede-Schumann, S. (2018). Apocynaceae. In: J. W. Kadereit & V. Bittrich (eds), *The Families and Genera of Vascular Plants, XV*: 207–208. Springer Verlag, Berlin & Heidelberg.

Endress, P. K., Davis, C. C. & Matthews, M. M. (2013). Advances in the floral structural characterization of the major subclades of Malpighiales, one of the largest orders of flowering plants. *Ann. Bot.* 111: 969–985.

Eriksen, B. & Persson, C. (2007). Polygalaceae. In: K. Kubitzki (ed.), *The Families and Genera of Vascular Plants, IX*: 345–363. Springer Verlag, Berlin & Heidelberg.

EUCLID *Eucalypts of Australia* (2020). 4th edition. https://apps.lucidcentral.org/euclid/text/intro/about.htm.

Fang, M.-Y., Fang, R.-C., He, M.-Y., Hu, L.-C., Yang, H.-P., Qin, H.-N., Min, T.-L., Chamberlain, D. F., Stevens, P. F., Wallace, G. D. & Anderberg, A. (2005). Ericaceae. In: Z. Y. Wu, P. H. Raven & D. Y. Hong (eds), *Flora of China* 14: 242–517. Science Press, Beijing & Missouri Botanical Garden Press, St. Louis.

Farjon, A. (2005). *A Monograph of Cupressaceae and* Sciadopitys. Royal Botanic Gardens, Kew.

Farjon, A. (2017). *A Handbook of the World's Conifers, Vol. 1*. 2nd edition (revised). Brill, Leiden & Boston.

Fischer, E. (2004). Scrophulariaceae. In: K. Kubitzki (ed.), *The Families and Genera of Vascular Plants, Vol. VII*: 333–432. Springer Verlag, Berlin & Heidelberg.

Flora of North America Editorial Committee (eds) (1993). *Flora of North America North of Mexico, Vol. 2*. Oxford University Press, New York & Oxford.

Flora of North America Editorial Committee (eds) (2006). *Flora of North America North of Mexico*. Oxford University Press, New York & Oxford. Vols 19, 20 & 21.

Freeman, C. C. (2016). Hydrangeaceae. In: Flora of North America Editorial Committee (eds), *Flora of North America North of Mexico* 12: 462–491. Oxford University Press, New York & Oxford.

Freeman, C. C., Rabeler, R. K. & Elisens, W. J. (2019a). Orobanchaceae. In: Flora of North America Editorial Committee (eds), *Flora of North America North of Mexico* 17: 465–768. Oxford University Press, New York & Oxford.

Freeman, C. C., Rabeler, R. K. & Elisens, W. J. (2019b). Plantaginaceae. In: Flora of North America Editorial Committee (eds), *Flora of North America North of Mexico* 17: 11. Oxford University Press, New York & Oxford.

Freeman, C. C. & Reveal, J. L. (2005). 44. Polygonaceae Jussieu, buckwheat family In: Flora of North America Editorial Committee (eds), *Flora of North America* 5: 405–456. Oxford University Press, New York & Oxford.

Friis, I. (1993). Urticaceae. In: K. Kubitzki, J. G. Rohwer & V. Bittrich (eds), *The Families and Genera of Vascular Plants, II*: 612–630. Springer Verlag, Berlin & Heidelberg.

Fryxell, P. A. (1997). The American genera of Malvaceae—II. *Brittonia* 49 (2): 204–269. New York Botanical Garden, New York.

Fu, D. Z. & Zhu, G. (2001). Ranunculaceae. In: Z. Y. Wu & P. H. Raven (eds), *Flora of China* 6: 133–438. Science Press, Beijing & Missouri Botanical Garden Press, St. Louis.

Funk, V. A., Susanna, A., Stuessy, T. F. & Bayer, R. J. (eds) (2009). *Systematics, Evolution, and Biogeography of Compositae*. International Association for Plant Taxonomy, Vienna; Smithsonian Institution, Washington, D.C.

Furlow, J. J. (1997). Betulaceae. In: Flora of North America Editorial Committee (eds), *Flora of North America North of Mexico* 3. Oxford University Press, New York & Oxford.

Galle, F. C. (1997). *Hollies. The Genus Ilex*. Timber Press, Inc., Portland.

George, L. O. (1997). *Podophyllum*. Berberidaceae. In: *Flora of North America* 3: 278–288. Oxford University Press, New York.

Goetghebeur, P. (1998). Cyperaceae. In: K. Kubitzki (ed.), *The Families and Genera of Vascular Plants, IV*: 141–190. Springer Verlag, Berlin & Heidelberg.

Goldblatt, P. & Manning, J. C. (2008). *The Iris Family*. Timber Press, Portland.

Goldblatt, P. (1993). Iridaceae. In: G. V. Pope (ed.), *Flora Zambesiaca* 12: 4. Royal Botanic Gardens, Kew.

Goldblatt, P. (1996). Iridaceae. In: R. M. Polhill (ed.), *Flora of Tropical East Africa*: 1–89. Balkema, Rotterdam.

Goldblatt, P. (1998). Iridaceae. In: K. Kubitzki (ed.), *The Families and Genera of Vascular Plants, III*: 295–333. Springer Verlag, Berlin & Heidelberg.

Govaerts, R., Simpson, D. A., Bruhl, J., Egorova, T., Goetghebeur, P. & Wilson, K. (2007). *World Checklist of Cyperaceae*. Royal Botanic Gardens, Kew.

Govaerts, R., Andrews, S., Coombes, A., Gilbert, M., Hunt, D., Nixon, D. & Thomas, M. (1998). *World Checklist of Fagaceae*. http://wcsp.science.kew.org. Facilitated by the Royal Botanic Gardens, Kew.

Govaerts, R., Frodin, D. G. & Radcliffe-Smith, A. (2000). *World Checklist and Bibliography of Euphorbiaceae (with Pandaceae)*. Royal Botanic Gardens, Kew.

Graham, S. A. & Graham, A. (2014). Ovary, fruit, and seed morphology of the Lythraceae. *Int. J. Pl. Sci.* 175: 202–240.

Graham, S. A., Hall, J., Sytsma, K. & Shi, S. H. (2005). Phylogenetic analysis of the Lythraceae based on four gene regions and morphology. *Int. J. Pl. Sci.* 166: 995–1017.

Graham, S., Diazgranados, M. & Barber, J. (2011). Relationships among the confounding genera *Ammannia*, *Hionanthera*, *Nesaea* and *Rotala* (Lythraceae). *Bot. J. Linn. Soc.* 166: 1–19.

Grant, V. (2003). Taxonomy of the Polemoniaceae: the subfamilies and tribes. *Sida* 20: 1371–1385.

Green, P. S. (2004). Oleaceae. In: K. Kubitzki (ed.), *The Families and Genera of Vascular Plants, Vol. VII*: 296–306. Springer Verlag, Berlin & Heidelberg.

Gu, C., Li, C., Lu, L., Jiang, S., Alexander, C., Bartholomew, B., Brach, A. R., Boufford, D. E., Ikeda, H., Ohba, H., Robertson, K. R. & Spongberg, S. A. (2003). Rosaceae. In: Z. Y. Wu, P. H. Raven & D. Y. Hong (eds), *Flora of China* 9: 46–434. Science Press, Beijing & Missouri Botanical Garden Press, St. Louis.

Hanes, M. M. (2015). Malvaceae. In: Flora of North America Editorial Committee (eds), *Flora of North America* 6: 187–375. Oxford University Press, New York & Oxford.

Harley, R. M., Atkins, S., Budantsev, A. L., Cantino, P. D., Conn, B. J., Grayer, R., Harley, M. M., de Kok, R. P. J., Krestovskaja, T., Morales, R., Paton, A. J., Ryding, O. & Upson, T. (2004). Labiatae. In: K. Kubitzki (ed.), *The Families and Genera of Vascular Plants, Vol. VII*: 167–275. Springer Verlag, Berlin & Heidelberg.

Hartmann, H. E. K. (1988). Fruit types in *Mesembryanthema. Beitr. Biol. Pflanzen* 63: 313–349.

Hartmann, H. E. K. (2017). *Illustrated Handbook of Succulent Plants. Aizoaceae, A–G; H–Z*. 2nd edition. Springer Verlag, Berlin.

Hauke, R. L. (1990). Equisetaceae. In: K. U. Kramer & P. S. Green (eds), *The Families and Genera of Vascular Plants, Vol. I*: 46–48. Springer Verlag, Berlin & Heidelberg.

Hedge, I. C. (1968). Cruciferae. In: K. H. Rechinger (ed.), *Flora Iranica: Flora des Iranischen Hochlandes und der Umrahmenden Gebirge, 57*. Akademische Druck und Verlagsanstalt, Graz.

Henderson, A. (2009). *Palms of Southern Asia*. New York Botanical Garden, New York; Princeton University Press, New Jersey.

Henderson, A., Galeano, G. & Bernal, R. (1995). *Field Guide to the Palms of the Americas*. Princeton University Press, New Jersey.

Herber, B. E. (2003). Thymelaeaceae. In: K. Kubitzki & C. Bayer (eds), *Families and Genera of Vascular Plants, V*: 373–396. Springer Verlag, Berlin & Heidelberg.

Herre, A. G. J. (1971). *The Genera of Mesembryanthemaceae*. Tafelberg, Cape Town.

Hevly, R. H. (1963). Adaptations of cheilanthoid ferns to desert environments. *J. Ariz. Acad. Sci.* 2 (4): 164–175.

Heywood, V. H. (1993). *Flowering Plants of the World*. B. T. Batsford Ltd, London.

Heywood, V. H. (2007a). Araliaceae. In: V. H. Heywood, R. K. Brummitt, A. Culham & O. Seberg (eds), *Flowering Plant Families of the World*. pp. 42–44. Royal Botanic Gardens, Kew.

Heywood, V. H. (2007b). Cistaceae. In: V. H. Heywood, R. K. Brummitt, A. Culham & O. Seberg (eds), *Flowering Plant Families of the World*. pp. 100–101. Royal Botanic Gardens, Kew.

Heywood, V. H., Brummitt, R. K., Culham, A. & Seberg, O. (2007). *Flowering Plant Families of the World*. Royal Botanic Gardens, Kew.

Ho, T.-N. & Pringle, J. S. (1995). Gentianaceae. *Flora of China* 16: 1–139. Science Press, Beijing; Missouri Botanical Garden Press, St. Louis.

Hodel, D. R. & Henrich, J. E. (2020). *Itoa orientalis*. PalmArbor 2020; 11: 1–29. https://ucanr.edu/sites/HodelPalmsTrees/files/334724.pdf

Hofmann, U. & Bittrich, V. (2016a). Caprifoliaceae. In: J. W. Kadereit & V. Bittrich (eds), *The Families and Genera of Vascular Plants, XIV*: 117–129. Springer Verlag, Berlin & Heidelberg.

Hofmann, U. & Bittrich, V. (2016b). Morinaceae. In: J. W. Kadereit & V. Bittrich (eds), *The Families and Genera of Vascular Plants, XIV*: 275–280. Springer Verlag, Berlin & Heidelberg.

Holmes, W. C. (2002). *Smilax*. In: Flora of North America Editorial Committee (eds), *Flora of North America* 26: 468. Oxford University Press, New York & Oxford.

Holttum, R. E. (1964). The tree-ferns of the genus *Cyathea* in Australasia and the Pacific. *Blumea* 12 (2): 241–274.

Huang, S. Kelly, L. M. & Gilbert, M. G. R. (2003). Aristolochiaceae. In: Z. Y. Wu, P. H. Raven & D. Y. Hong (eds), *Flora of China* 5: 246–269. Science Press, Beijing; Missouri Botanical Garden Press, St. Louis.

Huber, H. (1993). Aristolochiaceae. In: K. Kubitzki, J. G. Rohwer & V. Bittrich (eds), *The Families and Genera of Vascular Plants, II*: 129–137. Springer Verlag, Berlin & Heidelberg.

Hunt, D. (2016). *CITES Cactaceae Checklist*, 3rd edition. Royal Botanic Gardens, Kew.

Hunt, D. & Taylor, N. (2011). Cactaceae. In: J. Cullen, S. G. Knees & H. S. Cubey (eds), *The European Garden Flora, Vol. 2*, 2nd edition. Cambridge University Press, Cambridge.

Hunt, D., Taylor, N. & Charles, G. (2006). *The New Cactus Lexicon*. 2 vols (text + atlas). DH Books, Milborne Port.

Jeffrey, C. (1980). A review of the Cucurbitaceae. *Bot. J. Linn. Soc.* 81: 233–247.

Jessup, L. W. (1984). Celastraceae. *Flora of Australia* 22: 150–180.

Johnson, L. (2009). Polemoniaceae. Phlox family. In: *The Tree of Life Web Project*. http://tolweb.org/Polemoniaceae/

Johnson, O. & More, D. (2004). *Tree Guide*. Harper Collins Publishers, London.

Jones, D. L. (1988). *Native Orchids of Australia*. Reed Books, Frenchs Forest, New South Wales.

Judd, W. S., Campbell, C. S., Kellogg, E. A. & Stevens, P. F. (1999). *Plant Systematics, a Phylogenetic Approach*. Sinauer Associates, Sunderland,

Judd, W. S., Campbell, C. S., Kellogg, E. A., Stevens, P. F. & Donoghue, M. J. (2008). *Plant Systematics, a Phylogenetic Approach*. 3rd edition. Sinauer Associates, Sunderland.

Kadereit, J. W. (1993). Papaveraceae. In: K. Kubitzki, J. G. Rohwer & V. Bittrich (eds), *The Families and Genera of Vascular Plants, II*: 494–506. Springer Verlag, Berlin & Heidelberg.

Kalkman, C. (2004). Rosaceae. In: K. Kubitzki (ed.), *Families and Genera of Vascular Plants, VI*: 343–386. Springer Verlag, Berlin & Heidelberg.

Kellogg, E. A. (2015). *The Families and Genera of Vascular Plants, XIII. Poaceae*. Springer Verlag, Berlin.

Killick, D. J. B. (1976). Flacourtiaceae. In: J. H. Ross (ed.), *Flora of Southern Africa* 22: 53–92.

Kirschner, J., Balslev, H., Ceska, A., Coffey Swab, J., Edgar, E., Garcia-Herran, C., Hamet-Ahti, L., Kaplan, Z., Novara, L. J., Novikov, V. S. & Wilton, A. (2002a). Juncaceae 1: *Rostkovia* to *Luzula*. In: A. E. Orchard, J. Bleyerveen, A. J. G. Wilson & B. Kuchlmayr (eds), *Species Plantarum: Flora of the World, Part 6*. Australian Biological Resources Study, Canberra.

Kirschner, J., Brooks, R. E., Clemants, S. E., Ertter, B., Hamet-Ahti, L., Fernandez Carvajal Alvarez, M. C., Novara, L. J., Novikov, V. S., Simonov, S. S., Snogerup, S., Wilson, K. L. & Zika, P. F. (2002b). Juncaceae 2: *Juncus* subg. *Juncus*. In: A. E. Orchard, J. Bleyerveen, A. J. G. Wilson & B. Kuchlmayr (eds). *Species Plantarum: Flora of the World, Part 7*. Australian Biological Resources Study, Canberra.

Kirschner, J., Clemants, S. E., Ertter, B., Hamet-Ahti, L., Fernandez Carvajal Alvarez, M. C., Miyamoto, F., Novara, L. J., Novikov, V. S., Simonov, S. S., Snogerup, S. & Wilson, K. L. (2002c). Juncaceae 3: *Juncus* subg. *Agathryon*. In: A. E. Orchard, J. Bleyerveen, A. J. G. Wilson & B. Kuchlmayr (eds). *Species Plantarum: Flora of the World, Part 8*. Australian Biological Resources Study, Canberra.

Klak, C., Bruyns, P. V. & Hedderson, T. A. J. (2007). A phylogeny and new classification for Mesembryanthemoideae (Aizoaceae). *Taxon* 56: 737–756.

Klak, C., Hanáček, P. & Bruyns, P.V. (2017). Disentangling the Aizooideae: new generic concepts and a new subfamily in Aizoaceae. *Taxon* 66: 1147–1170.

Koch, M. A., German, D. A., Kiefer, M. & Franzke, A. (2018). Database taxonomics as key to modern plant biology. *Trends Pl. Sci.* 23 (1): 4–6.

Koehne, E. (1903). Lythraceae. In: A. Engler (ed.), *Das Pflanzenreich* IV: 1–326. Wilhelm Engelmann, Leipzig.

Köhler, E. (2009). Neotropical Buxaceae. In: W. Milliken, B. Klitgaard & A. Baracat, *Neotropikey—Interactive Key and Information Resources for Flowering Plants of the Neotropics.* http://www.kew.org/science/tropamerica/neotropikey/families/Buxaceae.htm.

Korall, P., Conant, D. S., Metzgar, J. S., Schneider, H. & Pryer, K. M. (2007). A molecular phylogeny of scaly tree ferns (Cyatheaceae). *Amer. J. Bot.* 94 (5): 873–886.

Koutroumpa, K., Theodoridis, S., Warren, B. H., Jiménez, A., Ferhat Celep, F., Doğan, M., Romeiras, M. M., Santos-Guerra, A., Fernández-Palacios, J. M., Caujapé-Castells, J., Moura, M., Menezes de Sequeira, M. & Conti, E. (2018). An expanded molecular phylogeny of Plumbaginaceae, with emphasis on *Limonium* (sea lavenders): taxonomic implications and biogeographic considerations. *Ecol. Evol.* 8 (24): 12397–12424.

Kubitzki, K. (1993a). Betulaceae. In: K. Kubitzki, J. G. Rohwer & V. Bittrich (eds), *The Families and Genera of Vascular Plants, II*: 152–156. Springer Verlag, Berlin & Heidelberg.

Kubitzki, K. (1993b) Cannabaceae. In: K. Kubitzki, J. G. Rohwer & V. Bittrich (eds), *The Families and Genera of Vascular Plants, II*: 204–206. Springer Verlag, Berlin & Heidelberg.

Kubitzki, K. (1993c). Fagaceae. In: K. Kubitzki, J. G. Rohwer & V. Bittrich (eds), *The Families and Genera of Vascular Plants, II*: 301–309. Springer Verlag, Berlin & Heidelberg.

Kubitzki, K. (1993d). Plumbaginaceae. In: K. Kubitzki, J. G. Rohwer & V. Bittrich (eds), *The Families and Genera of Vascular Plants, Plants, II*: 523–530. Springer Verlag, Berlin & Heidelberg.

Kubitzki, K. (2018). Apiaceae. In: K. Kubitzki & V. Bittrich (eds), *The Families and Genera of Flowering Plants, XV*: 9–206. Springer Verlag, Berlin & Heidelberg.

Kühn, R., Pedersen, H. Æ. & Cribb, P. (2019). *Field Guide to the Orchids of Europe and the Mediterranean.* Royal Botanic Gardens, Kew.

Kühn, U., Bittrich, V., Carolin, R., Freitag, H., Hedge, I. C., Uotila, P. & Wilson, P. G. (1993). Chenopodiaceae. In: K. Kubitzki, J. G. Rohwer & V. Bittrich (eds), *The Families and Genera of Vascular Plants, II*: 253–280. Springer Verlag, Berlin & Heidelberg.

Kunjun, F., Ohba, H. & Gilbert, M. G. (2001). Crassulaceae. In: Z. Y. Wu & P. H. Raven (eds), *Flora of China* 8: 202–268. Science Press, Beijing & Missouri Botanical Garden Press, St. Louis.

Lammers, T. G. (2007a). Campanulaceae. In: J. W. Kadereit & C. Jeffrey, *The Families and Genera of Vascular Plants, VIII*: 26–59. Springer Verlag, Berlin & Heidelberg.

Lammers, T. G. (2007b). *World Checklist and Bibliography of Campanulaceae.* Royal Botanic Gardens, Kew.

Large, M. F. & Braggins, J. E. (2009). *Tree Ferns.* Timber Press, Portland, Oregon.

Legume Phylogeny Working Group (LPWG). (2017). A new subfamily classification of the Leguminosae based on a taxonomically comprehensive phylogeny. *Taxon* 66 (1): 44–77.

Lewis, G., Schrire, B., Mackinder, B. & Lock, M. (eds) (2005). *Legumes of the World.* Royal Botanic Gardens, Kew.

Li, A, Bao Bojian, Grabovskaya-Borodina, A. E., Hong, S., McNeill, J., Mosyakin, S. L., Ohba, H. & Park, C. (2003). Polygonaceae. In: Z. Y. Wu, P. H. Raven, & D. Y. Hong (eds), *Flora of China* 5: 277–350. Science Press, Beijing; Missouri Botanical Garden Press, St. Louis.

Li, H. & Boyce, P. C. (2010). *Remusatia* (Araceae). *Flora of China* 23: 3–79. Science Press, Beijing; Missouri Botanical Garden Press, St. Louis.

Li, J. & Bogle, A. (2001). A new suprageneric classification system of the Hamamelidoideae based on morphology and sequences of nuclear and chloroplast DNA. *Harvard Pap. Bot.* 5: 499–515.

Li, P. & Skvortsov, A. K. (1999). Betulaceae. In: Z. Y. Wu & P. H. Raven (eds), *Flora of China* 4: 286–313. Science Press, Beijing; Missouri Botanical Garden Press, St. Louis.

Li, P., Leeuwenberg, A. J. M. & Middleton, D. J. (1995). Apocynaceae. *Flora of China* 16: 143–188. Science Press, Beijing; Missouri Botanical Garden Press, St. Louis.

Linder H. P., Briggs B. G. & Johnson L. A. S. (1998). Restionaceae. In: K. Kubitzki (ed.), *The Families and Genera of Vascular Plants, IV*: 425–445. Springer Verlag, Berlin & Heidelberg.

Linder, H. (1986). A review of the tropical African and Malagasy Restionaceae. *Kew Bull.* 41 (1): 99–106.

Linder, H. P. & Kurzweil, H. (1999). *Orchids of Southern Africa*. Balkema, Rotterdam.

Little, R. J. & McKinney, L. E. (2015). Violaceae. In: Flora of North America Editorial Committee (eds), *Flora of North America North of Mexico* 6: 106. [For *Cubelium concolor*, see *Hybanthus concolor*].

Liu, Q. & Zhou, L. (2008). Linaceae. In: Z. Y. Wu, P. H. Raven & D. Y. Hong (eds), *Flora of China* 11: 34–38. Science Press, Beijing; Missouri Botanical Garden Press, St. Louis.

Loconte, H. (1993). Berberidaceae. In: K. Kubitzki, J. G. Rohwer & V. Bittrich (eds), *The Families and Genera of Vascular Plants, II*: 147–152. Springer Verlag, Berlin & Heidelberg.

Lowrie, A. (1987–1998). *Carnivorous Plants of Australia, Vols 1–3*. University of Western Australia Press, Nedlands.

Lowry, P. P. II & Plunkett, G. M. (2020). Resurrection of the genus *Heptapleurum* for the Asian clade of species previously included in *Schefflera* (Araliaceae). *Novon* 28: 143–170.

Lu, A., Stone, D. E. & Grauke, L. J. (1999). Juglandaceae. In: Z. Y. Wu & P. H. Raven (eds), *Flora of China* 4: 277–285. Science Press, Beijing; Missouri Botanical Garden Press, St. Louis.

Lu, D., Wu, Z., Zhou, L., Chen, S., Gilbert, M. G., Lidén, M., McNeill, J., Morton, J. K., Oxelman, B., Rabeler, R. K., Thulin, M., Turland, N. J. & Wagner, W. L. (2001). Caryophyllaceae. In: Z. Y. Wu, P. H. Raven & D. Y. Hong (eds), *Flora of China* 6: 1–113. Science Press, Beijing; Missouri Botanical Garden Press, St. Louis.

Ma, J.-S., Ball, P. W. & Levin, G. A. (2016). Celastraceae. In: Flora of North America Editorial Committee (eds), *Flora of North America* 12: 111–132. Oxford University Press, New York & Oxford.

Ma, J.-S., Zhang, Z.-X., Liu, Q.-R., Peng, H. & Funston, A. M. (2008). In: Z. Y. Wu, P. H. Raven & D. Y. Hong (eds), *Flora of China* 11: 439–492. Science Press, Beijing; Missouri Botanical Garden Press, St. Louis.

Magnolia Society. (2021). www.magnoliasociety.org/.

Majeed, A., Singh, A., Choudhary, S. & Bhardwaj, P. (2018). RNAseq-based phylogenetic reconstruction of Taxaceae and Cephalotaxaceae. *Cladistics* 35: 461–468.

Manchester, S. R. (1989). Systematics and fossil history of the Ulmaceae. In: P. R. Crane & S. Blackmore (eds), *Evolution, Systematics and the Fossil History of the Hamamelidae, Vol. 2 "Higher" Hamamelidae*. pp. 221–251. Oxford University Press, Oxford.

Manning, J. (2007). *Field Guide to Fynbos*: 246–274. Struik Nature, Cape Town.

Manos, P. S., Cannon, C. H. & Oh, S-H. (2008). Phylogenetic relationships and taxonomic status of the paleoendemic Fagaceae of western North America: recognition of a new genus, *Notholithocarpus*. *Madroño* 55: 181–190.

Manos, P.S. & Stone, D.E. (2001). Evolution, systematics and phylogeny of the Juglandaceae. *Ann. Missouri Bot. Gard.* 88 (2): 231–269.

Marais, W. & Verdoorn, I. C. (1963). Gentianaceae. In: R. A. Dyer, L. E. Codd & H. B. Rycroft (eds), *Flora of Southern Africa* 26: 171–243. Government Printer, Pretoria.

Matthews, M. L. & Endress, P. K. (2011). Comparative floral structure and systematics in Rhizophoraceae, Erythroxylaceae and the potentially related Ctenolophonaceae, Linaceae, Irvingiaceae and Caryocaraceae (Malpighiales). *Bot. J. Linn. Soc.* 166: 331–416.

Mayer, V. (2016). Dipsacaceae. In: J. W. Kadereit & V. Bittrich (eds), *The Families and Genera of Vascular Plants, XIV*: 145–163. Springer Verlag, Berlin & Heidelberg.

Mayo, S. J., Bogner, J. & Boyce, P. C. (1997). *The genera of Araceae*. Royal Botanic Gardens, Kew.

McDill, J., Repplinger, M., Simpson, B. B. & Kadereit, J. W. (2009). The phylogeny of *Linum* and Linaceae subfamily Linoideae, with implications for their systematics, biogeography, and evolution of heterostyly. *Syst. Bot.* 34: 386–405.

Medan, D. & Schirarend, C. (2004). Rhamnaceae. In: K. Kubitzki (ed.), *The Families and Genera of Vascular Plants, VI*: 320–338. Springer Verlag, Berlin & Heidelberg.

Meerow, A. W. (2005). *Betrock's Cold Hardy Palms*. Betrock Information Systems, Inc., Davie, Florida.

Meney, K. A. & Pate, J. (1999). *Australian Rushes: Biology, Identification and Conservation of Restionaceae and Allied Families*. UWA Publishing, Perth.

Min, T.-L. & Bartholomew, B. (2007). Theaceae. *Flora of China* 12: 366–478. Science Press, Beijing; Missouri Botanical Garden Press, St. Louis.

Mitchell, A. D., Frodin, D. G. & Heads, M. J. (1997). Reinstatement of *Raukaua*, a genus of the Araliaceae centred in New Zealand. *New Zealand J. Bot.* 35: 309–315.

Mitchell, J. (1990). The poisonous Anacardiaceae genera of the World. *Advances in Economic Botany* 8: 103–129. New York Botanical Garden Press, New York.

Moore, D. M. (1976). Dipsacaceae. In: T. G. Tutin, V. H. Heywood, N. A. Burges, D. M. Moore, D. H. Valentine, S. M. Walters & D. A. Webb, *Flora Europaea* 4: 56–74. Cambridge University Press, Cambridge.

Moore, L. B. & Edgar, E. (1970). *Flora of New Zealand. II. Indigenous Tracheophyta: Monocotyledones except Gramineae*. Botany Division DSIR, Wellington.

Moran, R. V. (2009). Crassulaceae. In: Flora of North America Editorial Committee (eds), *Flora North America North of Mexico* 8: 147. Oxford University Press, New York & Oxford.

Murrell, Z. E. & Poindexter, D. B. (2016). Cornaceae. In: Flora of North America Editorial Committee (eds), *Flora of North America* 12: 443–458. Oxford University Press, New York & Oxford.

Nandi, O. I. (1998). Floral development and systematics of Cistaceae. *Pl. Syst. Evol.* 212: 107–134.

Navas Bustamante, L. E. (1973). *Flora de la Cuenca de Santiago de Chile* 1: 149. Comisión Central de Publicaciones de la Universidad de Chile, Santiago.

Nesom, G. L., Schmidt, C. L. & Wilken, D. H. (2016). Rhamnaceae. In: Flora of North America Editorial Committee (eds), *Flora of North America* 12: 43. Oxford University Press, New York & Oxford.

Nicolas, A. N. & Plunkett, G. M. (2009). The demise of subfamily Hydrocotyloideae (Apiaceae) and the re-alignment of its genera across the entire order Apiales. *Molec. Phylogenet. Evol.* 53: 134–151.

Nixon, K. C. (1989). Origins of Fagaceae. In: P. R. Crane & S. Blackmore (eds), *Evolution, Systematics and the Fossil History of the Hamamelidae, Vol.* 2. pp. 23–43. Oxford University Press, Oxford.

Nordenstam, B. (1998). Colchicaceae. In: K. Kubitzki (ed.), *The Families and Genera of Vascular Plants, III*: 175–185. Springer Verlag, Berlin & Heidelberg.

Oliver, E. G. H. (2012). Ericaceae. In: J. C. Manning & P. Goldblatt (eds), *Plants of the Greater Cape Floristic Region, Vol. 1: The Core Cape Flora* 29. pp. 482–511. South African National Biodiversity Institute (SANBI Publishing), Pretoria.

Ooststroom, S. J. van. (1953). Convolvulaceae. *Flora Malesiana, ser. I*, 4: 388–512; 5: 558–564.

Oxelman, B., Kornhall, P., Olmstead, R. G. & Bremer, B. (2005). Further disintegration of Scrophulariaceae. *Taxon* 54: 411–425.

Pastore, J. F. B. (2018). *Polygala veadeiroensis* (Polygalaceae), a new species of *Polygala* endemic to Chapada dos Veadeiros, Goiás, Brazil. *Kew Bull.* 73: 37.

Pastore, J. F. B., Abbott, J. R., Neubig, K. M., Whitten, W. M., Mascarenhas, R. B., Mota, M. C. A. & van den Berg, C. (2017). A molecular phylogeny and taxonomic notes in *Caamembeca* (Polygalaceae). *Syst. Bot.* 42: 54–62.

PBI Solanum Project. (2021). Solanaceae Source. www.solanaceaesource.org/.

Perkins, A. J. (2020). Araliaceae, In: P. G. Kodela (ed.), *Flora of Australia*. Australian Biological Resources Study, Canberra.

Phipps, J. B. (2015). Rosaceae. In: Flora of North America Editorial Committee, *Flora of North America* 9. Oxford University Press, New York & Oxford.

Plants of the World Online (POWO) (2019 onwards). http://www.plantsoftheworldonline.org/. Facilitated by the Royal Botanic Gardens, Kew.

Plunkett, G. M., Wen, J., Lowry, P. P., Mitchell, A. D., Henwood, M. J. & Fiaschi, P. (2018). Araliaceae. In: J. W. Kadereit & V. Bittrich (eds), *The Families and Genera of Vascular Plants, XV*: 413–446. Springer Verlag, Berlin & Heidelberg.

Porter, J. M. & Johnson, L. A. (2000). A phylogenetic classification of Polemoniaceae. *Aliso* 19: 55–91.

Pridgeon, A. M., Cribb, P. J., Chase, M. W. & Rasmussen, F. N. (eds) (1999–2014). *Genera Orchidacearum. Vols 1–6*. Oxford University Press, Oxford.

Prince, L. M. (2007). A brief nomenclatural review of genera and tribes in Theaceae. *Aliso* 24: 105–121.

Prince, L. M. (2009). Theaceae. In: Flora of North America Editorial Committee, *Flora of North America North of Mexico* 8: 322–328. Oxford University Press, New York & Oxford.

Pryer, K. M., Schneider, H., Smith, A. R., Cranfill, R., Wolf, P. G., Hunt, J. S. & Sipes, S. D. (2001). Horsetails and ferns are a monophyletic group and the closest living relatives to seed plants. *Nature* 409 (6820): 618–622.

Purdie, R. W., Symon, D. E. & Haegi, L. (1982). Solanaceae. *Flora of Australia* 29. Australian Government Publishing Service, Canberra.

Qiner, Y., Deyuan, H., Malecot, V., Bouffard, D. E. (2011a). Adoxaceae. In: Z. Y. Wu, P. H. Raven & D. Y. Hong (eds), *Flora of China* 19: 570–614. Science Press, Beijing; Missouri Botanical Garden Press, St. Louis.

Qiner, Y., Landrein, S., Osborne, J. & Borosova, R. (2011b). Caprifoliaceae. In: Z. Y. Wu, P. H. Raven & D. Y. Hong (eds), *Flora of China* 19: 616–641. Science Press, Beijing; Missouri Botanical Garden Press, St. Louis.

Qiuyan, X. & Boufford, D. E. (2005). Cornaceae. In: Z. Y. Wu, P. H. Raven & D. Y. Hong (eds), *Flora of China* 14: 206–221. Science Press, Beijing; Missouri Botanical Garden Press, St. Louis.

Rabeler, R. K. & Hartman, R. L. (2005). Caryophyllaceae. In: Flora of North America Editorial Committee (eds), *Flora of North America North of Mexico* 5: 3–215. Oxford University Press, New York & Oxford.

Rabeler, R. K., Freeman, C. C. & Elisens, W. J. (2019). Scrophulariaceae. In: Flora of North America Editorial Committee (eds), *Flora of North America North of Mexico* 17: 324–325. Oxford University Press, New York & Oxford.

Radcliffe-Smith, A. (2001). *Genera Euphorbiacearum*. Royal Botanic Gardens, Kew.

Raven, P. H. (1988). Onagraceae as a model of plant evolution. In: L. D. Gottlieb & S. K. Jain (eds), *Plant Evolutionary Biology: A symposium Honoring G. Ledyard Stebbins*. pp. 85–107. Chapman & Hall, London.

Rhui-cheng, F. & Wilken, D. H. (1995). Polemoniaceae. *Flora of China* 16: 326–327. Science Press, Beijing; Missouri Botanical Garden Press, St. Louis.

Rogers, Z. S. (2009). *A World Checklist of Thymelaeaceae (version 1)*. Missouri Botanical Garden, St. Louis.

Rohwer, J. G. (1993). Lauraceae. In: K. Kubitzki, J. G. Rohwer & V. Bittrich (eds), *The Families and Genera of Vascular Plants, II*: 366–391. Springer Verlag, Berlin & Heidelberg.

Rohwer, J. G. & Berg, C. C. (1993). Moraceae. In: K. Kubitzki, J. G. Rohwer & V. Bittrich (eds), *The Families and Genera of Vascular Plants, II*: 438–453. Springer Verlag, Berlin & Heidelberg.

Särkinen, T., Bohs, L., Olmstead, R. G. & Knapp, S. (2013). A phylogenetic framework for evolutionary study of the nightshades (Solanaceae): a dated 1000-tip tree. *BMC Evol. Biol.* 13: 214.

Schieber, A., Stintzing, F. & Carle, R. (2001). By-products of plant food processing as a source of functional compounds—recent developments. *Trends Food Sci. Technol.* 12: 401–413.

Schuettpelz, E., Schneider, H., Huiet, L., Windham, M. D. & Pryer, K. M. (2007). A molecular phylogeny of the fern family Pteridaceae: assessing overall relationships and the affinities of previously unsampled genera. *Molec. Phylogenet. Evol.* 44 (3): 1172–1185.

Schwartsburd, P. B., Perrie, L. R., Brownsey, P., Shepherd, L. D., Shang, H., Barrington, D. S. & Sundue, M. A. (2020). New insights into the evolution of the fern family Dennstaedtiaceae from an expanded molecular phylogeny and morphological analysis. *Molec. Phylogenet. Evol.* 150: 106881.

Semmouri, I., Bauters, K., Léveillé-Bourret, É., Starr, J. R., Goetghebeur, P. & Larridon, I. (2019). Phylogeny and systematics of Cyperaceae, the evolution and importance of embryo morphology. *Bot. Rev.* 85: 1–39.

Shang, H., Sundue, M., Wei, R., Wei, X. P., Luo, J. J., Liu, L., Schwartsburd, P. B., Yan, Y. H. & Zhang, X. C. (2018). *Hiya*: a new genus segregated from *Hypolepis* in the fern family Dennstaedtiaceae, based on phylogenetic evidence and character evolution. *Molec. Phylogenet. Evol.* 127: 449–458.

Sherman-Broyles, S. L., Barker, W. T. & Schulz, L. M. (1997). Ulmaceae mirbel—elm family. In: Flora of North America Editorial Committee (eds), *Flora of North America North of Mexico* 3: 368–380. Oxford University Press, New York & Oxford.

Shi, Z., Chen, Y., Chen, Y., Ling, Y-R., Liu, S., Ge, X., Gao, G., Zhu, S., Liu, Y., Humphries, C. J., Yang, C., Raab-Straube, E., Gilbert, M. G., Nordenstam, B., Kilian, N., Brouillet, L., Illarionova, I. D., Hind, D. J. N., Jeffrey, C., Bayer, R. J., Kirschner, J., Greuter, W., Anderberg, A. A., Semple, J. C., Štěpánek, J., Freire, S. E., Martins, L., Koyama, H., Kawahara, T., Vincent, L., Sukhorukov, A. P., Mavrodiev, E. V. & Gottschlich, G. (2011). Asteraceae (Compositae). In: Z. Y. Wu, P. H. Raven & D. Y. Hong (eds), *Flora of China* 20–21: 1–894. Science Press, Beijing; Missouri Botanical Garden Press, St. Louis.

Simmons, M. P. (2004). Celastraceae. In: K. Kubitzki (ed.), *The Families and Genera of Vascular Plants, VI*: 29–64. Springer Verlag, Berlin & Heidelberg.

Simões, A. R. & Staples, G. W. (2017). Dissolution of tribe Merremieae and a new classification for its constituent genera. *Bot. J. Linn. Soc.* 183: 561–586.

Simpson, D. A. & Inglis, C. A. (2001). Cyperaceae of economic, ethnobotanical and horticultural importance: a checklist. *Kew Bull.* 56: 257–360.

Simpson, M. G. (2010). *Plant Systematics.* 2nd edition. Academic Press, Oxford.

Simpson, M. G. (2019). *Plant Systematics.* 3rd edition. Academic Press, Oxford.

Sleumer, H. (1977). Revision der Gattung *Azara* R. & P. (Flacourtiaceae). Mit 1 Abbildung im Text. *Bot. Jahrb. Syst.* 98 (2): 151–175.

Soltis, D. E. (2007). Saxifragaceae. In: K. Kubitzki (ed.), *The Families and Genera of Vascular Plants, IX*: 418–435. Springer Verlag, Berlin & Heidelberg.

Soza, V. L. & Olmstead, R. G. (2010). Molecular systematics of tribe Rubieae (Rubiaceae): evolution of major clades, development of leaf-like whorls, and biogeography. *Taxon* 59 (3): 755–771.

Ståhl, B. & Anderberg, A. A. (2004). Myrsinaceae. In: K. Kubitzki (ed.), *The Families and Genera of Vascular Plants, VI*: 266–281. Springer Verlag, Berlin & Heidelberg.

Stefanović, S., Austin, D. F. & Olmstead, R. G. (2003). Classification of Convolvulaceae: a phylogenetic approach. *Syst. Bot.* 28 (4): 791–806.

Stevens, P. F. (2001 onwards). *Angiosperm Phylogeny Website.* www.mobot.org/MOBOT/research/APweb/.

Stevens, P. F., Luteyn, J., Oliver, E. G. H., Bell, T. L., Brown, E. A., Crowden, R. K., George, A. S., Jordan, G. J., Ladd, P., Lemson, K., Mclean, C. B., Menadue, Y., Pate, J. S., Stace, H. M. & Weiller, C. M. (2004a). Ericaceae. In: K. Kubitzki (ed.), *The Families and Genera of Vascular Plants, VI*: 145–194. Springer Verlag, Berlin & Heidelberg.

Stevens, P. F., Dressler, S. & Weitzman, A. L. (2004b). Theaceae. In: K. Kubitzki (ed.), *The Families and Genera of Vascular Plants, VI*: 463–471. Springer Verlag, Berlin & Heidelberg.

Stone, D. E. (1993). Juglandaceae. In: K. Kubitzki, J. G. Rohwer & V. Bittrich (eds), *The Families and Genera of Vascular Plants, II*: 348–357. Springer Verlag, Berlin & Heidelberg.

Struwe, L. & Pringle, J. S. (2018). Gentianaceae. In: J. W. Kadereit & V. Bittrich (eds), *The Families and Genera of Vascular Plants, XV*: 453–503. Springer Verlag, Berlin & Heidelberg.

Sytsma, K. J., Morawetz, J., Pires, J. C., Nepokroeff, M., Conti, E., Zjhra, M. & Chase, M. W. (2002). Urticalean rosids: circumscription, rosid ancestry, and phylogenetics based on *rbcL, trnL-F*, and *ndhF* sequences. *Amer. J. Bot.* 89 (9): 1531–1546.

Tamura, M. (1993). Ranunculaceae. In: K. Kubitzki, J. G. Rohwer & V. Bittrich (eds), *The Families and Genera of Vascular Plants, II*: 563–583. Springer Verlag, Berlin & Heidelberg.

Tamura, M. N. (1998a). Liliaceae. In: K. Kubitzki (ed.), *The Families and Genera of Vascular Plants, III*: 343–353. Springer Verlag, Berlin & Heidelberg.

Tamura, M. N. (1998b). Melanthiaceae. In: K. Kubitzki (ed.), *The Families and Genera of Vascular Plants, III*: 369–380. Springer Verlag, Berlin & Heidelberg.

Tank, D. C., Beardsley, P. M., Kelchner, S. A. & Olmstead, R. G. (2006). L. A. S. Johnson Review 7. Review of the systematics of Scrophulariaceae s.l. and their current disposition. *Austral. Syst. Bot.* 19: 289–307.

Taylor, N. & Zappi, D. (2009 onwards). Neotropical Cucurbitaceae. In: W. Milliken, B. Klitgaard & A. Baracat, *Neotropikey—Interactive Key and Information Resources for Flowering Plants of the Neotropics.* www.kew.org/science/tropamerica/neotropikey/families/Cucurbitaceae.htm.

Thiesen, I. & Fischer, E. (2004). Myoporaceae. In: K. Kubitzki (ed.), *The Families and Genera of Vascular Plants, Vol. VII*: 289–292. Springer Verlag, Berlin & Heidelberg.

Thompson, S. A. (2000). Araceae. In: Flora of North America Editorial Committee, *Flora of North America* 22. Oxford University Press, New York & Oxford.

Thornhill, A. H., Ho, S. Y., Külheim, C. & Crisp, M. D. (2015). Interpreting the modern distribution of Myrtaceae using a dated molecular phylogeny. *Molec. Phylogenet. Evol.* 93: 29–43.

Townsend, C. C. (1993). Amaranthaceae. In: K. Kubitzki, J. G. Rohwer & V. Bittrich (eds), *The Families and Genera of Vascular Plants, II*: 70–91. Springer Verlag, Berlin & Heidelberg.

Trias-Blasi, A. T., Parnell, J. A. N., Pornpongrungrueng, P. & Kochaipat, P. (2020). Vitaceae. In: K. Chayamarit & H. Balslev (eds), *Flora of Thailand* 14 (4): 588–669.

Trias-Blasi, A., Suksathan, P. & Tamura, M. N. (2017). Melanthiaceae. In: T. Santisuk, *Flora of Thailand* 13 (3): 520–524. The Forest Herbarium, Bangkok.

Tryon, R. M. (1990). Pteridaceae. In: K. U. Kramer & P. S. Green (eds), *The Families and Genera of Vascular Plants, I*: 230–256. Springer Verlag, Berlin & Heidelberg.

Tucker, G. (2009) Ericaceae. In: Flora of North America Editorial Committee (eds), *Flora of North America* 8: 370–535. Oxford University Press, New York.

Tutin, T. G. (1968). Rhamnaceae. In: T. G. Tutin, V. H. Heywood, N. A. Burges, D. M. Moore, D. H. Valentine, S. M. Walters & D. A. Webb (eds), *Flora Europaea* 2: 243–245. Cambridge University Press, Cambridge & New York.

Tutin, T. G., Heywood, V. H., Burges, N. A., Moore, D. M. & Valentine, D. H. (eds) (1976). *Flora Europaea 4. Plantaginaceae to Compositae (and Rubiaceae).* Cambridge University Press, Cambridge & New York.

Tutin, T. G., Heywood, V. H., Burges, N. A., Moore, D. M., Valentine, D. H., Walters, S. M., Webb, D. A., Chater, A. O. & Richardson, I. B. K. (eds) (1980). *Flora Europaea. 5: Alismataceae to Orchidaceae (Monocotyledones).* Cambridge University Press, Cambridge & New York.

Tutin, T. G., Pritchard, N. M. & Melderis, A. (1972). Gentianaceae. In: T. G. Tutin, V. H. Heywood, N. A. Burges, D. M. Moore, D. H. Valentine, S. M. Walters & D. A. Webb (eds), *Flora Europaea* 3: 56–67. Cambridge University Press, Cambridge & New York.

Tutin, T. G., Heywood, V. H., Burges, N. A., Moore, D. M., Valentine, D. H., Walters, S. M. & Webb, D. A. (eds) (1972). Plumbaginaceae. In: T. G. Tutin, V. H. Heywood, N. A. Burges, D. M. Moore, D. H. Valentine, S. M. Walters & D. A. Webb (eds), *Flora Europaea* 3: 29–51. Cambridge University Press, Cambridge & New York.

Utteridge, T. (2015). Violaceae. In: T. Utteridge & G. Bramley (eds) (2015). *The Kew Tropical Plant Families Identification Handbook*. Royal Botanic Gardens, Kew.

Utteridge, T. & Bramley, G. (eds) (2020). *The Kew Tropical Plant Families Identification Handbook*. 2nd edition (revised). Royal Botanic Gardens, Kew.

Valentine, D. H. & Chater, A. O. (1972). Boraginaceae. In: T. G. Tutin, V. H. Heywood, N. A. Burges, D. M. Moore, D. H. Valentine, S. M. Walters & D. A. Webb (eds), *Flora Europaea* 3: 83–122. Cambridge University Press, Cambridge & New York.

van der Werff, H. (1991). A key to the genera of Lauraceae in the New World. *Ann. Missouri Bot. Gard.* 78: 377–387.

van der Werff, H. (2001). An annotated key to the genera of Lauraceae in the Flora Malesiana region. *Blumea* 46: 125–140.

van der Werff, H. & Richter, H. G. (1996). Toward an improved classification of Lauraceae. *Ann. Missouri Bot. Gard.* 83: 409–418.

Various contributors. (1978–ongoing). *Advances in Legume Systematics*, Royal Botanic Gardens, Kew.

Verdcourt, B. (1996). Lauraceae. *Flora of Tropical East Africa*. A. A. Balkema, Rotterdam.

Wagner, W. L., Hoch, P. C. & Raven, P. H. (2007). Revised classification of the Onagraceae. *Syst. Bot. Monogr.* 83. The American Society of Plant Taxonomists.

Wahlert, G. A., Marcussen, T., de Paula-Souza, J., Feng, M. & Ballard, H. E., Jr. (2014). A phylogeny of the Violaceae (Malpighiales) inferred from plastid DNA sequences: implications for generic diversity and intrafamilial taxonomy. *Syst. Bot.* 39 (1): 239–252.

Wallander, E. & Albert, V. A. (2000). A phylogeny and classification of Oleaceae based on *rps16* and *trnL-trn-F* sequence data. *Amer. J. Bot.* 87: 1827–1841.

Wallander, E. (2014). *The Oleaceae Information Site*. www.oleaceae.info/

Walters, S. M. (1976a). Caprifoliaceae. In: T. G. Tutin, V. H. Heywood, N. A. Burges, D. M. Moore, D. H. Valentine, S. M. Walters & D. A. Webb. *Flora Europaea* 4: 44–48. Cambridge University Press, Cambridge.

Walters, S. M. (1976b). Valerianaceae. In: T. G. Tutin, V. H. Heywood, N. A. Burges, D. M. Moore, D. H. Valentine, S. M. Walters & D. A. Webb. *Flora Europaea* 4: 48–56. Cambridge University Press, Cambridge.

Walters, S. M. (1993). Caryophyllaceae. In: T. G. Tutin, N. A. Burges, A. O. Chater, J. R. Edmondson, V. H. Heywood, D. M. Moore, D. H. Valentine, S. M. Walters & D. A. Webb (eds), *Flora Europaea* 1: 139–246. 2nd edition. Cambridge University Press, Cambridge.

Watson, L. & Dallwitz, M. J. (1992 onwards). *The Families of Flowering Plants: Descriptions, Illustrations, Identification and Information Retrieval*. www1.biologie.uni-hamburg.de/b-online/delta/angio/index.htm

Webb, D. A. (1972). Scrophulariaceae. In: T. G. Tutin, V. H. Heywood, N. A. Burges, D. M. Moore, D. H. Valentine, S. M. Walters & D. A. Webb (eds), *Flora Europaea* 3: 202–281. Cambridge University Press, Cambridge & New York.

Weberling, F. & Bittrich, V. (2016). Valerianaceae. In: J. W. Kadereit & V. Bittrich (eds), *The Families and Genera of Vascular Plants, XIV*: 385–401. Springer Verlag, Berlin & Heidelberg.

Webster, G. L. (1994). Classification of the Euphorbiaceae; synopsis of the genera and suprageneric taxa of Euphorbiaceae. *Ann. Missouri Bot. Gard.* 81 (1): 33–144.

Weddell, H. A. (1869). Urticaceae. In: A. De Candolle (ed.), *Prodromus Systematis Naturalis Regni Vegetabilis* 16 (1): 32–235. Paris.

Well, E. F. & Elvander, P. E. (2009). Saxifragaceae. In: Flora of North America Editorial Committee (eds), *Flora North America North of Mexico* 8: 43. Oxford University Press, New York and Oxford.

Wen, J., Lu, L.-M., Nie, Z.-L., Liu, X.-Q., Zhang, N., Ickert-Bond, S., Gerrath, J., Manchester, S. R., Boggan, J. & Chen, Z-D. (2018). A new phylogenetic tribal classification of the grape family (Vitaceae). *J. Syst. Evol.* 56: 262–272.

Weston P. H. (2007). Proteaceae. In: K. Kubitzki (ed.), *The Families and Genera of Vascular Plants, IX*: 364–404. Springer Verlag, Berlin & Heidelberg.

Wilken, D. H. (2004). Polemoniaceae. In: K. Kubitzki (ed.), *The Families and Genera of Vascular Plants, VI*: 300–312. Springer Verlag, Berlin & Heidelberg.

Wilkin, P. (2015a). Amaryllidaceae. In: T. M. A. Utteridge & G. Bramley (eds), *The Kew Tropical Plant Families Identification Handbook*. 1st edition. pp. 34–35. Royal Botanic Gardens, Kew.

Wilkin, P. (2015b). Asparagaceae. In: T. M. A. Utteridge & G. Bramley (eds), *The Kew Tropical Plant Families Identification Handbook*. 1st edition. pp. 36–37. Royal Botanic Gardens, Kew.

Wilmot-Dear, C. M. (2015a). Cannabaceae. In: T. M. A. Utteridge & G. Bramley (eds), *The Kew Tropical Plant Families Identification Handbook*. 1st edition. pp. 74–75. Royal Botanic Gardens, Kew.

Wilmot-Dear, C. M. (2015b). Moraceae. In: T. M. A. Utteridge & G. Bramley (eds), *The Kew Tropical Plant Families Identification Handbook*. 1st edition. pp. 76–77. Royal Botanic Gardens, Kew.

Wilmot-Dear, C. M. (2015c). Urticaceae. In: T. M. A. Utteridge & G. Bramley (eds), *The Kew Tropical Plant Families Identification Handbook*. 1st edition. pp. 78–79. Royal Botanic Gardens, Kew.

Wilmot-Dear, C. M. & Friis, I. (2013). The Old World species of *Boehmeria* (Urticaceae, tribus Boehmerieae). A taxonomic revision. *Blumea* 58: 85–216.

Wilson, P. G. (2011). Myrtaceae. In: K. Kubitzki (ed.), *The Families and Genera of Vascular Plants, X*: 212–271. Springer Verlag, Berlin & Heidelberg.

Wolf, P. G., Rowe, C. A., Kinosian, S. P., Der, J. P., Lockhart, P. J., Shepherd, L. D., McLenachan, P. A. & Thomson, J. A. (2019). Worldwide relationships in the fern genus *Pteridium* (bracken) based on nuclear genome markers. *Amer. J. Bot.* 106 (10): 1365–1376.

Wood, J. R. I., Williams, B. R. M., Mitchell, T. C., Carine, M. A., Harris, D. J. & Scotland, R. W. (2015). A monograph of *Convolvulus* L. (Convolvulaceae). *Phytokeys* 51: 1–282.

World Checklist of Vascular Plants (2021). version 2.0. http://wcvp.science.kew.org/ Facilitated by the Royal Botanic Gardens, Kew.

Wu, Z. Y., Monro, A. K., Milne, R. I., Wang, H., Yi, T. S., Liu, J. & Li, D. Z. (2013a). Molecular phylogeny of the nettle family (Urticaceae) inferred from multiple loci of three genomes and extensive generic sampling. *Molec. Phylogenet. Evol.* 69: 814–827.

Wu, Z. Y., Raven, P. H. & Hong, D. Y. (eds) (2007). *Flora of China* 12. Science Press, Beijing; Missouri Botanical Garden Press, St. Louis.

Wu, Z., Raven, P. H. & Hong D. Y. (eds) (2013b). *Flora of China Vols 2–3 (Lycophytes & Ferns)*. Science Press, Beijing; Missouri Botanical Garden Press, St. Louis.

Wurdack, K. J., Hoffmann, P. & Chase, M. W. (2005). Molecular phylogenetic analysis of uniovulate Euphorbiaceae (Euphorbiaceae *sensu stricto*) using plastid *rbcL* and *trnL-F* DNA sequences. *Amer. J. Bot.* 92 (8): 1397–1420.

Wurdack, K. J. & Horn, J. W. (2001). A re-evaluation of the affinities of the Tepuianthaceae: molecular and morphological evidence for placement in the Malvales. http://2001.botanyconference.org/section12/abstracts/264.shtml

Xu Langran & Aedo, C. (2008). Geraniaceae. In: Z. Y. Wu, P. H. Raven & D. Y. Hong (eds), *Flora of China* 11: 8–31. Science Press, Beijing; Missouri Botanical Garden Press, St. Louis.

Yang, J., Boufford, D. & Brach, A. R. (2011). Berberidaceae. In: Z. Y. Wu & P. H. Raven (eds), *Flora China* 19: 714–800. Science Press, Beijing; Missouri Botanical Garden Press, St. Louis.

Yang, Q. & Zmarzty, S. (2007). Flacourtiaceae. In: Z. Y. Wu, P. H. Raven & D. Y. Hong (eds), *Flora of China* 13: 112–137. Science Press, Beijing; Missouri Botanical Garden Press, St. Louis.

Zhang, L. & Turland, N. J. (2013). Equisetaceae. In: Z. Y. Wu, P. H. Raven & D. Y. Hong (eds), *Flora of China* 2–3: 67–72. Science Press, Beijing; Missouri Botanical Garden Press, St. Louis.

Zhang, M., Su, Z., Liden, M. & Grey-Wilson, C. (2008). Papaveraceae. In: Z. Y. Wu & P. H. Raven (eds), *Flora of China* 7: 261–428. Science Press, Beijing; Missouri Botanical Garden Press, St. Louis.

Zhang, Z., Lu, A. & D'Arcy, W. G. (1994). Solanaceae. In: Z. Y. Wu & P. H. Raven (eds), *Flora of China* 17: 300–332. Science Press, Beijing; Missouri Botanical Garden Press, St. Louis.

Zhao, F., Chen, Y-P., Salmaki, Y., Drew, B. T., Wilson, T. C., Scheen, A-C., Cerlep, F., Brauchler, C., Bendiksby, M., Wang, Q., Min, D-Z., Peng, H., Olmstead, R. G., Li, B. & Xiang, C-L. (2021). An updated tribal classification of Lamiaceae based on plastome phylogenetics. *BMC Biology* 19: 2.

Zhu Gelin, Mosyakin, S. L. & Clemants, S. E. (2003). Chenopodiaceae. In: Z. Y. Wu, P. H. Raven & D. Y. Hong (eds), *Flora of China* 5: 351–414. Science Press, Beijing; Missouri Botanical Garden Press, St. Louis.

Zomlefer, W. B., Judd, W. S., Whitten, W. M. & Williams, N. H. (2006). A synopsis of Melanthiaceae (Liliales) with focus on character evolution in tribe Melanthieae. *Aliso* 566–578.

Index of plant names

Featured plant families in **bold**, other families mentioned in the text in roman. Page number for main account of featured family in **bold**.